George Vivian Poore

Essays on Rural Hygiene

George Vivian Poore

Essays on Rural Hygiene

ISBN/EAN: 9783337854119

Printed in Europe, USA, Canada, Australia, Japan

Cover: Foto ©berggeist007 / pixelio.de

More available books at **www.hansebooks.com**

ESSAYS

ON

RURAL HYGIENE

BY

GEORGE VIVIAN POORE, M.D., F.R.C.P.

'Ces tas d'ordures du coin des bornes, ces tombereaux de boue cahotés la nuit dans les rues, ces affreux tonneaux de la voirie, ces fétides écoulements de fange souterraine que le pavé vous cache, savez-vous ce que c'est? C'est de la prairie en fleur, c'est de l'herbe verte, c'est du serpolet et du thym et de la sauge, c'est du gibier, c'est du bétail, c'est le mugissement satisfait des grands bœufs le soir, c'est du foin parfumé, c'est du blé doré, c'est du pain sur votre table, c'est du sang chaud dans vos veines, c'est de la santé, c'est de la joie, c'est de la vie. Ainsi le veut cette création mystérieuse qui est la transformation sur la terre et la transfiguration dans le ciel.

'Rendez cela au grand creuset; votre abondance en sortira. La nutrition des plaines fait la nourriture des hommes.

'Vous êtes maîtres de perdre cette richesse, et de me trouver ridicule par-dessus le marché. Ce sera là le chef-d'œuvre de votre ignorance.'

VICTOR HUGO (*Les Misérables*)

LONDON
LONGMANS, GREEN, AND CO.
AND NEW YORK: 15 EAST 16th STREET
1893

PREFACE

EIGHT of the thirteen chapters of this work have, in whole or in part, been previously published. Chapter II. formed the subject of a 'free lecture' delivered at University College, London, in 1892, and was subsequently published in the *Lancet*. Chapter III., 'On the Shortcomings of Some Modern Sanitary Methods,' was delivered as an address to the Sanitary Institute at its annual meeting in 1887. This chapter has been translated into German, and has been published by Wagner of Graz with the title, *Die Nachtheile einiger neueren sanitären Methoden*. Chapter IV., 'The Living Earth,' was written for an address delivered at Brighton in 1891, when the author held the office of President of the section of Preventive Medicine at the Health Congress held in that town.

Chapters V., VI., VII., and XII. formed part of a handbook written in 1884 at the request of the Council of the Health Exhibition, which was held in London in that year, and they are incorporated in this work by arrangement with Messrs. W. Clowes & Sons, who own the copyright of the Health Exhibition Handbooks.

That part of Chapter VII. which deals with the subject of 'Hygienic Units' was contributed to the 'Medical Magazine' in 1892.

Chapter XIII., 'The Story of Bremontier,' formed the subject of an address delivered in York in 1886 during a Health Congress held in that city under the auspices of the Sanitary Institute.

Chapters I., VIII., IX., X., and XI. have not been previously published.

Although the work has been written in a somewhat desultory manner during the past nine years, the author has always had in view the publication, as a whole, of the parts of which it is composed. The reader will find that in each chapter a different set of facts is marshalled in illustration of the central idea, and that the amount of overlapping of the chapters is not more than in any case would have been inevitable. The repetition of an important fact is often advisable, and doubly advisable when such repetition is for the purpose of bringing it into new relations.

The thanks of the author are due to his friend Dr. BUTLER HARRIS, who has kindly revised the proof-sheets and prepared the index.

30 WIMPOLE STREET, W.: *March* 1893.

CONTENTS

CHAPTER	PAGE
I. INTRODUCTORY	1
II. THE CONCENTRATION OF POPULATION IN CITIES	10
III. ON THE SHORTCOMINGS OF SOME MODERN SANITARY METHODS	49
IV. 'THE LIVING EARTH'	85
V. THE HOUSE	119
VI. AIR	142
VII. WATER	157
VIII. PRACTICAL DETAILS	192
IX. PERSONAL EXPERIENCES IN A COUNTRY TOWN	219
X. PERSONAL EXPERIENCES (*continued*):— WATER-SUPPLY	235
XI. PERSONAL EXPERIENCES IN A LONDON SUBURB	255
XII. BURIAL	284
XIII. THE STORY OF BREMONTIER	297
APPENDIX	315
INDEX	319

ILLUSTRATIONS

FIG.		PAGE
1. Conical Metal Vessel	. . . .	. 112
2. Glass Funnels	. . .	. 115
3. Gooseberries and Raspberries		
4. Gooseberries	.	*Between pages* 230
5. Black and White Currants		*and* 231
6. Peas	. . .	

ESSAYS
ON
RURAL HYGIENE

CHAPTER I
INTRODUCTORY

IN the ensuing chapters the author will attempt to show that many of the hygienic arrangements which have been in vogue for some years are largely based upon erroneous scientific principles; and are, therefore, bad from other points of view, political, moral, economic, and hygienic.

The title of 'Rural Hygiene' has been chosen because it is only in places having a rural or semi-rural character that it is possible to be guided by scientific principles in our measures for the preservation of health and the prevention of disease. In cities the hygienic arrangements are the products of expediency rather than principle, and are not unfrequently carried out in defiance of the teachings of pure science. Overcrowding is encouraged, and rivers or other sources of water are recklessly fouled, because such conditions are, or are supposed to be, 'good for trade.' Our municipal governors, who are mainly selected from the trading classes, and the majority of whom have had no scientific training of any kind, are rarely capable of looking beyond the question of immediate profit, which to them seems all-important. If a so-called sanitary measure seems likely to increase the rateable value of a district for the time being, that is generally regarded as sufficient ground for action, and money is recklessly borrowed to carry out expensive and half-considered measures, in which the work of sanitation is merely begun and rarely completed.

The hygienic measures of cities have, for the most part, been hastily adopted in order to escape the dangers which are inseparable from an undue concentration of population. They may be compared to the herculean method which was practised upon the stables of King Augeas, and although we may admire the prowess of Hercules, we can feel nothing but contempt for Augeas, who would have been happier and richer had he kept his oxen in a rational way.

The 'good for trade' doctrine is fatal to sound sanitary measures, because Mr. 'Good for Trade' (as Bunyan might have called him) will not for a moment listen to any proposition for limiting, however slightly, the concentration of population. 'Good for trade' is fond of big schemes, the money for which is borrowed, and so long as the money be spent he does not much care upon what. There can be no doubt that the almost unlimited powers of borrowing which our municipalities possess have been harmful in so far as they have encouraged sanitary authorities to act precipitately when otherwise they would have been obliged to move slowly and cautiously, and in doing so have gained their experience.

Mr. 'Good for Trade' is often not deficient in cunning, and is quite capable of seeing how advantageous are big sanitary schemes for landowners, building speculators, water shareholders, contractors, and the crowd of tradesmen who follow in their train. He thoroughly understands the various methods by which what is known as 'bringing down the Local Government Board' is brought about, and then, when the groaning ratepayer complains, he protests that the local authority is helpless because of the action of the Government.

The title of 'Rural Hygiene' has been chosen not only because it is in the country alone that Hygiene can be based upon *principles* rather than *expediency*, but also from a feeling that if the rural element be entirely banished from our towns, and if the fearful concentration of population which is seen in the modern city, both here and in America, be allowed to proceed unchecked we are in a fair way to increase rather than decrease the liability of our towns to suffer from epidemics.

'God made the country and man made the town,' said the poet Cowper, and this is a saying which is not only true but filled with deep meaning for those who are interested in the physical and moral welfare of our populations. It is to be hoped that rural districts will make every effort to retain their rural character, and that, if those sanitary measures which foster overcrowding be forced upon them, they will put some compensating restraint upon the owner of 'eligible sites' and the speculative builder. With our modern methods of communication such dangerous concentration of population cannot be necessary, and one hopes that before the nineteenth century closes people will begin to see the advantages not only of the *rus in urbe* but also of the *urbs in rure*.

We hear a great deal of the dulness of rural districts, but it must be remembered that in reality our modern methods of communication have placed all the more solid advantages of the town within reach of the villages, and that a dweller in even the most remote of our villages is able, if he be so minded, at very small expense, to keep himself abreast of modern ideas in all departments of knowledge.

The big towns are daily becoming more and more a menace to the country. Might is right in the present day as it ever has been, and it makes little difference whether oppression be practised by fists or votes.

If, however, big towns be allowed to devastate rural districts and drain rivers to dryness in order that they may have a gigantic water-supply with the main object of still further concentrating the population in their boundaries; if the big towns be allowed to use the neighbouring counties as dumping-grounds for nuisances; if they be allowed to foul the rivers so that the fishing industries are destroyed, and it is dangerous for the countryman's cattle to drink from them; if they be allowed to empty their filth along the foreshore of places which are miles distant; if they be permitted to send their paupers into one county, their lunatics into a second, their scarlet fever cases to a third, and their small-pox to a fourth, it is evident that the country people must rush for the towns in self-defence. It is to be hoped that it may some day be recognised that rural districts have their rights. Free trade is indispensable for the existence of our big towns, and while it has enabled the country at large to grow rich it certainly has not been an unmixed blessing to rural districts. The rural districts have, so to say, sacrificed themselves for the sake of the towns, and are now finding out that they have been nurturing a set of Frankenstein's monsters, which are devoid of gratitude and return evil for good.

In the chapters which follow an attempt will be made

to set forth the true principles which should guide our actions in procuring a healthy house, pure air, good water, and cheap and wholesome food.

It will be first of all demonstrated that a neglect of sound principles lands us in difficulties, and accordingly the first of the subsequent chapters will be devoted to the evils which arise from an over-concentration of population; it will further be insisted upon that this over-concentration is an indirect effect of our modern sanitary methods, which give what may be called a fatal facility for the packing of houses in dangerous proximity to each other. This over-concentration is often defended on the ground that it is necessary for the purposes of commerce, but not a few facts will be brought forward which tend to throw considerable doubts on the economic advantages of such concentration. It will be argued that the retention of a rural element in our rapidly developing towns and the allowing of open spaces to inter-penetrate and dovetail with the houses has not only great advantages on the score of health, but will be shown to be equally advantageous when regarded in its purely financial and economic aspects.

In the third chapter the shortcomings of modern sanitary methods will be dealt with, and it will be shown that the mixing of putrescible matter with water is a fundamental scientific error which leads to the dissemination of water-borne diseases, the pollution of rivers, and the poisoning of wells. Whether such methods be regarded by the modern light of bacteriology or by the evils and expenses of which they are notoriously the cause, they must be condemned as unscientific, thriftless, and immoral. They are unscientific because they encourage putrefaction and hinder nitrification, which should be our aim in dealing with organic refuse; they are thriftless because they merely waste or practically destroy that which, rightly used, should be a source of profit and productiveness; and they are immoral because, by merely 'passing on' our refuse to be a nuisance elsewhere than on our own premises, we show a forgetfulness of our duty towards our neighbour and we do unto others that which we are unwilling that others should do unto us.

In the fourth chapter, on 'The Living Earth,' it will be shown that the humus possesses (in virtue of the animal and vegetable organisms which it contains) a marvellous power not only of turning organic matter into food for plants by what is known as the process of nitrification, but that while in this way it tends to increase our food supplies, it is no less powerful, if rightly and scientifically used, to protect our wells from all dangerous animal pollutions.

In dealing with the subject of 'The House' in the fifth chapter, the many evils which are practically inseparable from what are known as modern sanitary fittings are passed in review, and it is insisted upon that no house can be securely and permanently wholesome unless it have tolerably direct relations with cultivable land. The modern practice of erecting houses on insufficient area and the evil consequences resulting from such practices are dwelt upon, and it is pointed out that the planning of town houses involves problems which need not and ought not to trouble the architect of country houses. A few words will be said on the subject of those buildings which are designed for the reception of large numbers of people, such as hotels, barracks, and schools.

The subject of 'Air' is discussed in the sixth chapter. A few elementary facts with regard to air are first given, and then the relationship which exists between the earth and the air is discussed; it is insisted upon that vegetation is essential for the freshening of the air ; and that when, as in overcrowded cities, the air becomes too foul to allow vegetation to flourish, this in itself constitutes a danger to health, for without vegetation neither is the air freshened nor the soil purified.

The point which receives most attention when discussing the question of 'Water' in the seventh chapter is the great difficulty of purifying water which has once been fouled, and the doubts which surround all the common processes which are at present advocated for that purpose. It will be shown that, if we want pure water, a scientific and careful bestowal of putrescible refuse is the first thing necessary; that while putrescible matter mixed with water and allowed to accumulate in underground receptacles not only escapes the salutary action of the humus, but, leaking under pressure, inevitably trickles unchanged to our wells, however deep they may be, the same matter, if superficially buried in the humus and allowed to oxidise and nitrify, is not likely to foul a properly made well, no matter how shallow it may be. The necessary relations which exist between earth and water are discussed, and it is shown that every individual requires a definite minimum amount of earth, air, and water in order to live. Modern methods of communication combined with the unrestricted importations of food have enabled us to neglect what may be called the 'earth-unit,' and to concentrate our population in a disastrous fashion, and it is largely due to this neglect of the earth-unit that our increasing difficulties with regard to municipal water-supplies are due, difficulties which must, one would fear, grow progressively greater.

In the eighth, ninth, tenth, and eleventh chapters the author gives his personal experiences in dealing with the problem of domestic sanitation. The principles which should guide us in the management of sanitary details are discussed, and arrangements are described which are believed to be satisfactory, at least in the circumstances for which they were designed. It must, however, always be borne in mind that circumstances cannot be neglected when dealing with questions of sanitation, and that the discovery of a sanitary panacea will certainly never be made. The peculiar questions which arise when dealing with the sanitation of cottages will be discussed, and the best methods of refuse-disposal and water-supply in relation to cottages will receive attention. The burning sanitary questions which arise in country towns and in growing suburbs will be illustrated by personal experiences, which the bulk of readers will recognise as having been in no degree exceptional.

A chapter on 'Burial' has been inserted as being part of the great question of the right bestowal of effete organic matter, and the power of the earth to deal with the dead body as satisfactorily as it deals with all other forms of dead organic matter is insisted upon.

The final chapter is devoted to the story of Bremontier, which is inserted as the best example known to the author of the glorious results which have been obtained by a patient waiting upon nature in the true scientific spirit.

It is hardly necessary for the author to say that in bringing out this little book he has been prompted solely by a deep sense of the importance of the subject which is discussed in it. It is important not only to the individual but to the nation at large. The bestowal of refuse is a problem which confronts every individual daily and almost hourly. We may practically destroy it so that our native soil gets no advantage, or we may start it upon a round of creative productiveness which will provide food, warmth, houses, and raiment, and in so doing find perpetual occupation for the increasing numbers of 'the unemployed.' To what extent it might be possible for the towns to mitigate the prevailing agricultural distress by supplying farmers with manurial matters at a cheap rate is a question well worthy of consideration. Looked at philosophically, the question of the right bestowal of organic refuse is a national question of great political importance which no statesman can neglect, and which has effects vastly more far-reaching than is generally supposed. It is hoped that every patriotic man will ponder the question seriously, and will recognise that it has its moral side. It is the duty of each of us to take care that we do not, by apathetic carelessness or culpable ignorance, endanger the health of others, and we must remember that it is no excuse for the adoption of bad and dangerous methods of sanitation to urge that they are 'convenient.' Such an excuse might be put forward in defence of acts which we all recognise as criminal. We are individually under a moral obligation to see the refuse of our dwellings safely bestowed, so as not to endanger the health of others; but in these matters we are too prone to allow 'rates' to take the place of morals, and to expect a collection of individuals, merely because they are called a 'board,' to be able to do for us that which we ignorantly profess to be unable to do for ourselves. In sanitary matters, more than in any others perhaps, we are gradually losing our freedom; but it is high time for the individual to rouse himself to a sense of duty, and insist on his right to individual liberty.

CHAPTER II

THE CONCENTRATION OF POPULATION IN CITIES

THE steadily increasing tendency of population to leave the country and concentrate in towns is a fact which does not admit of a doubt. Of the twenty-nine millions of inhabitants of England and Wales, about 18,500,0C0 are classified by the Registrar-General as belonging to town districts and about 10,500,000 as belonging to country districts. In 1801 London contained about one-eleventh of the entire population of England and Wales, whereas, according to the last census, it was found to contain about one-seventh of the entire population. Dr. Gould, of Washington, in a paper read before the seventh International Congress of Hygiene and Demography, stated that a short half-century ago the urban population of the United States was 8·5 per cent. of the whole, while to-day the urban population constitutes 29 per cent. of the whole. The American city, says Dr. Gould, 'creates itself with appalling suddenness,' and it is probable that Chicago, with its 1,200,000 inhabitants, having doubled its population in the last ten years, may be said in this particular matter to have beaten the record. It must not be forgotten, however, that London has more than doubled its population in the last half century; that Cardiff has risen from 80,000 to 130,000 inhabitants in the last ten years; that

Barrow-in-Furness, Eastbourne, Bournemouth, West Ham, and Croydon are all instances of towns, besides many others, which have sprung into existence within the memory of the present generation. It is important to bear these facts in mind, as showing that the English cities, equally with those in America, create themselves with appalling suddenness; that we do not merely inherit our cities with their various shortcomings, but that we create them for ourselves, and are directly and solely answerable for their good or ill construction.

The cause of this concentration of population is the desire for business, the wish to get money easily and quickly, and to spend it advantageously. A man who trades in a centre of express mail services, telegraphs, and telephones has all the world before him if he knows how to make use of it. There is consequently a rush for such favoured centres, and it has come about that steam and electricity, which annihilate time and space, instead of enabling us to live further apart from each other, have produced a directly opposite effect. The dealer, be he wholesale or retail, likes to be surrounded by a crowd of potential customers, rather than be dependent on a few; and the artisan naturally turns to great industrial centres as offering the readiest market for his labour, and often finds out too late that the higher wages of the town are more than counterbalanced by the extra cost of living. The crowds of independent and idle persons who settle in the towns do so because they find a greater variety of methods of killing time, or, as they prefer to put it, because the state of civilisation is greater in the towns than in the country. But what is civilisation? A recent anonymous writer (the author of 'Behind the Bungalow') speaks of this 'half-hatched civilisation of ours, which merely distracts our energies by multiplying our needs, and leaves us no better off than we were before we discovered them'; and we must all admit that there is a good deal of this kind of civilisation which passes current at the present day, especially in cities, where a large proportion of the population are the slaves of inane conventionalities.

Cities are the abodes of art, and art in all its forms is elevating; but it is a question whether the increasing difficulties which, especially in London, we encounter in the study and appreciation of nature do not more than counterbalance the artistic advantages. 'Pictures, taste, Shakespeare, and the musical-glasses' are not the only things worthy the attention of civilised man, and it is noteworthy that many of the leading spirits in all ages have turned from the artificial enjoyments of the town to the greater freedom and more natural pleasures of the country. It is said that our rural fellow-subjects, having been forcibly educated to a pitch which enables them to study all the gay doings of the town, are beginning to find the country insufferably dull. The cause of dulness, however, is in ourselves and not in our surroundings. Our country friends must be taught that they have at hand one of the surest cures for dulness. Let them learn to study and appreciate the book of nature—that book which is always with us, ever open and inexhaustible—and dulness will become impossible.

The mind which has occupation to really interest it cannot be dull or weary, and the surest way to find interest in this life is productive labour. In the making of a pudding, the raising of crops, the writing of books, the practice of a profession, and in all the constructive arts, it is necessary to use the judgment; and it is this exercise of judgment which is the true complement of book-learning. The man who reads and who produces nothing is seldom capable of action; he becomes giddy with the opinions of others, and finds it impossible to have a fixed opinion of his own. In action and production the mere see-saw of criticism is of no use, we must make up our minds. We must really think, and the man who has learnt to think runs little risk of being dull.

Those who have to cope with the uncertainties of nature are bound to exercise their judgment in a high degree. Unless the farmer be constantly thinking and looking ahead, and unless he bring (often unconsciously) a good deal of science to bear upon his practical work, he will have no chance of success. We are not to sit in judgment on such a one because he does not show the alacrity, sharpness, and power of repartee which are seen in the town dweller. The cockney and the yokel are educated in totally different schools, neither understands the other, and they have for each other a good deal of mutual contempt. It is difficult, however, to believe that he who can perform the varied duties of the farm, and who has been brought up in contact with a wide range of natural phenomena, is not the equal intellectually of the factory hand, who is the slave of a machine which thunders to and fro with brutal accuracy for ten hours at a stretch. Physically, the country labourer is vastly superior to the town dweller, and be it remembered that mental power is mainly dependent on physical health.

One of the undoubted consequences of the concentration of population in towns is deterioration in physical health. The disease-rate and the death-rate are both higher in urban than in country districts. The difficulties of rearing children are much greater in towns than in the country, and the risk of faulty development in those which are reared is also greater. The undue concentration of population must tend towards the deterioration of our race, and there can be no doubt that the diminution of the evils of such concentration is a subject which demands very earnest attention. We all of us have to run more or less risk in earning our living. Some enter the army or navy, some go to tropical countries, some become file-grinders, hotel servants, or doctors, and it is not, perhaps, desirable that the bread-winners should look too narrowly at the risks connected with their daily work. Every prudent man, however, will endeavour to secure vigorous health for his wife and children; and, whenever it can be avoided, these should not be called upon to run unnecessary risks.

The Registrar-General's returns show that for every 100 deaths in country districts there are about 120 deaths in towns, and that the death-rate of children under five, per 1,000 living at that age, was (in the year 1890) 66·5 in Lancashire and 65·2 in London, as against 30·3 in Dorsetshire, 31·1 in Wiltshire, 31·8 in Berkshire, 31·2 in Rutlandshire, 35·8 in Herefordshire, 35·3 in Shropshire, 35·8 in Hampshire, 36·4 in Oxfordshire, 37·5 in Bedfordshire, 36·0 in Sussex, and 39·3 in Somersetshire; and, although the infant mortality was higher in the other counties having a more urban character, in none of them (except Lancashire and London) did it exceed 57·5. It is thus evident that the risk to children under five is twice as great in London and Lancashire as it is in some of the rural districts. How many children are crippled for life by infantile diseases contracted in towns we have no means of knowing. Death-rates are nowadays the subject of perennial paragraphs in the newspapers, they are systematically used for puffing localities, and are beginning to form a recognised feature in auctioneers' advertisements.

Before accepting the death-rate as evidence of the healthiness of a locality we must take several facts into consideration. 1. The mobility of the population in the present day is greater than it ever has been previously. This must have a vitiating effect upon the value of the death-rate, for it is evident to all of us that a man who has contracted a fatal disease, let us say in London, may die at Croydon, Brighton, the Riviera, Cairo, or elsewhere. This fact must largely lessen the trustworthiness of local death-rates, although it does not, probably, appreciably affect the trustworthiness of the death-rate for the country as a whole. 2. Again, the death-rate of a city is of little value unless correction be made for abnormal age distribution. London, especially the central parts of it, contains a great deficiency of persons at the extreme (and most vulnerable) periods of life, and this, of course, helps to keep the death-rate lower than it would be if it contained its due proportion of tender infants and feeble old persons. The population of central London, it must be remembered, is largely composed of selected adults imported from the country. 3. Lastly, let it be observed that the death-rate of London is kept down very largely by a process of dilution. If any comparison is to be made between the London of to-day and the London of former times, we must be careful to select identical areas. As it is, the high death-rate of the centre is diluted by the low death-rates of the outskirts, and the healthiness of Hampstead, Lewisham, and other outlying districts is used to conceal the condition of the centre.

For the sake of his argument the author has divided the population of London into three portions as nearly equal as the very irregular shape of the registration districts will permit.

1. The outer portion, consisting of the outlying districts of Hammersmith, Fulham, Hampstead, Islington, Hackney, Poplar, Battersea, Wandsworth, Camberwell, Lewisham, and Plumstead, and containing 1,676,475 inhabitants according to the last census.

2. Districts holding a middle position—viz. Kensington, Chelsea, Paddington, St. George's (Hanover Square), Marylebone, St. Pancras, Lambeth, Greenwich, and Woolwich, and containing a population of 1,317,078 inhabitants.

3. The central districts, which might be spoken of as London proper. These include, in addition to the 'central districts' of the Registrar-General, Westminster, Shoreditch, Bethnal Green, Whitechapel, St. George's-in-the-East, Stepney, Mile End, Southwark, Newington, Bermondsey, and Rotherhithe. These districts form a compact city in a ring-fence, and contain 1,217,503 inhabitants, a population twice as big as that of any other city in the country. The deaths of the various

| London Population 4,211,056 | Death-rates per 1,000 living ||||||||||||
| | 1890 |||| 1891 |||| 1892 ||||
	1st Qtr.	2nd Qtr.	3rd Qtr.	4th Qtr.	1st Qtr.	2nd Qtr.	3rd Qtr.	4th Qtr.	1st Qtr.	2nd Qtr.	3rd Qtr.	4th Qtr.
Outlying Districts, Population 1,676,475	20·9	15·4	16·5	21	20·4	20·2	14·1	17·6	25·7	16·1	15·5	16·8
Middle Districts, Population 1,317,078	24·3	17·6	17·6	22	24·3	23·3	16·8	19·7	27·4	18·5	15·7	17·5
Central Districts, Population 1,217,503	29·6	21	22·0	28·8	28	27·9	20·7	24·9	32·0	22·3	20·2	21·2

districts as published in the corrected quarterly tables of the Registrar-General have been taken, and from them death-rates for these three divisions of London have been calculated for every quarter during the last three years.

If we take the yearly rates instead of the quarterly we have the following result :—

	1890	1891	1892
Outlying districts	18·4 per 1,000	18·1 per 1,000	18·5 per 1,000
Middle districts	20·4 ,,	21·0 ,,	19·7 ,,
Central districts	25·6 ,,	25·4 ,,	23·9 ,,

From the above tables it is evident that when we boast of London as a healthy city we must exclude the central districts, or London proper, with its 1,200,000 inhabitants. Indeed, the death-rate of London proper was extremely high in these three years; but, large as the figures are, they do not show the full state of the case, because no correction has been made for age distribution and mobility, which is essential before we can compare them with the rates for the rest of the country. Nothing can show better than these figures the effect of overcrowding on death-rate. Not only are the central districts the most crowded and therefore the poorest, but they are hemmed in on every side by the middle and outlying districts, so that fresh air can never reach them. As a result, we find the death-rates are some 40 per cent. higher than the rates for the outlying districts. In all seasons and under all conditions the uniformity of relation between the rates of the outlying and central districts is maintained. If we take the first quarter of 1890, the second quarter of 1891, and the first quarter of 1892, during each of which we had an epidemic of influenza, we find that the death-rate of the outlying districts was 20·9, 20·2, and 25·7, while for the central districts it was 29·6, 27·9, and 32·0. Or, take the last quarter of 1890 and the first quarter of 1891, which were characterised by weather of great severity, we find that in the outlying districts the death-rate was 21 and 20·4, while in the central districts it was 28·8 and 28, notwithstanding the fact that the temperature of the central districts was certainly considerably higher than that of the outlying districts. If we take the periods when the climatic conditions are at their best, the same relationship in the figures is maintained. Thus, for the third quarter of the three years the rates of the outlying districts were 16·5, 14·1, and 15·5, while for the central districts they were 22·9, 20·7, and 20·2.

Finally, let us take a period when the climatic conditions were at their worst, as during the fog of Christmas 1891, which commenced on December 21, and continued for parts of five days. During this fog the mean temperature of the air was 25° F. ; the air was calm, with slight easterly currents, and saturated with moisture ; ozone and sunshine absent. Under these circumstances the irritating products of combustion and the soot from thousands of chimneys remained suspended in the fog, as did also the effluvia from over 4,000,000 human beings and the varied impurities of a big city. The air was bitterly cold and irritating, and from the absence of wind could not be renewed, so that the citizens of London were very much in the condition of gold fish in a bowl whose water has not been changed. The corrected average number of deaths for the fortnight ending January 2, 1892, was 3,728, while the deaths which actually occurred were 5,170, so that we have 1,442 deaths caused by the fog. The death-rate for the whole of London for this fortnight was 32, the figures for the three areas being 29 for the outlying districts, 31·8 for the middle districts, and 38 for the central districts. This high death-rate for the central districts, which are the warmest, seems to show that it is not merely the coldness of the fog which raises the death-rates, but rather the impurities, mechanical, chemical and infective, which it contains.

Bad as are the health conditions in the centre of London, they are eclipsed by those of New York. In the city of New York, which had a population of 1,600,000 in 1890, more than 1,200,000 of the inhabitants live in tenement houses. The density of population averages 65 persons to the acre, while in many of the wards it is over 200 to the acre, and in one ward it reached 559 persons to the acre. The average death-rate for the past ten years was over 26, and in the year 1890 the deaths exceeded the births by 980; and while there were 39,250 births in the year, the authorities are jubilant because only 10,288 of the children born, or rather more than 26 per cent., died before they were a year old. The deaths in New York invariably exceed the births, so that the population is maintained and increased entirely by immigrants. The birth-rate is exceedingly low. Last year it was only 24, and that was a higher figure than had been reached in any of the thirty previous years. It must be remembered that New York has a magnificent water-supply and a very complete system of sewerage. As in London, there is apparently a very great deficiency of very young and very old inhabitants; and again, as in London, it may be presumed that a very large number of those who fall ill retire from the city to die elsewhere. Notwithstanding the great activity of the sanitary officers, it is evident that the state of public health in New York is far from good, and that, owing mainly to the density of population, it is not a very wholesome place.

The evils of overcrowding have been made more plain and comprehensible by the discoveries which have been made in the new science of bacteriology. When the public hears that a new microbe has been discovered in connection with this or that disease, it says: 'Now, surely we shall have a cure.' It may be well to point out that, hitherto, in no case of general disease has the discovery of a microbe led to the discovery of a cure; and the evidence that the study of the life-history of pathogenic microbes is likely to lead to the discovery of 'cures' is, to say the least, not very strong. 'Prevention,' however, is better than 'cure'; and it is clear that the study of the life-history of pathogenic microbes must precede all adequate measures for preventing the diseases which they cause.

It was the study of the life-history of the microbes which produce putrefaction which led Sir Joseph Lister to adopt the antiseptic treatment of wounds which has produced such grand results. The knowledge that cholera and typhoid poisons can live in water has led to great caution in the selection and protection of water-supplies, and it is possible that the time is not far distant when we shall regard the wilful mixing of excremental matters with water as, in a sense, an act of public suicide. Again, the knowledge that typhoid fever, diphtheria, and scarlet fever are often connected with the milk-supply has been the means of arresting many an epidemic. In short, the medical profession has no cause to feel ashamed of its recent progress in the direction of prevention of disease, which, in the words of Sir William Jenner, should be the great aim of a physician.

While certain diseases reach us *viâ* water and food,

there are others which come to us in the air we breathe, and it is well to bear in mind that it is in relation to air-borne diseases that overcrowding is especially dangerous. Whooping-cough, measles, scarlet fever, diphtheria, tubercle, typhus fever, small-pox, influenza, pneumonia, and plague, and, to a less extent, typhoid fever, are among the diseases which are air-borne, and the spread of which is, in a very obvious way, favoured by overcrowding. It is definitely known with regard to some, and may be safely inferred with regard to all the above-named diseases, that the carriers of the infection are particulate and alive. A sufferer from any one of these diseases is continually giving off infective particles, and the danger of the infection travelling to a second person depends (1) upon the distance of the second from the first ; (2) the ease with which the infected air can be diluted with fresh air. Thus the danger caused by proximity to an infected person in the open air is comparatively slight, because the chance that the infective particles will be blown away by the wind is very great. When, however, the infected and the healthy come to live under the same roof, to occupy the same room, and, still more, the same bed, the risks of infection are enormously increased, because the air in the confined space gets very largely charged with infective particles which the healthy can hardly avoid taking. Given the rate of movement of the air, the degree of proximity of the individuals, and the number of infective particles given off by the diseased person, and the risk of infection could be stated as a mathematical formula. There is good reason to believe, however, that the mere fact of overcrowding produces some change (possibly chemical) which greatly increases the risks of infection, apart from the mathematical factors above alluded to. Some of the facts connected with the recent epidemic of influenza bring out very clearly the increased danger of infection which results from overcrowding between walls and under a common roof.

Overcrowding is seen at its maximum in the forecastle of a ship after the men have 'turned in,' and the danger to all should any case of infective disease be present among them is well known. The explosive outbursts of influenza which are liable to occur on board ship have been amongst the startling features of every epidemic. In the same way explosive outbursts of the disease have occurred in establishments where many persons have a common employment under a common roof, as witness the sudden occurrence of a thousand cases at the end of 1889 in the Magazins du Louvre at Paris, and a similar outburst at the General Post Office in January 1890. A careful consideration of the whole of the facts of the epidemic of 1889-90 leads Dr. Parsons, in his report made to the Local Government Board, to conclude:[1] " 8. That in public services and establishments persons employed together in large numbers in enclosed spaces have suffered in larger proportion than those employed few together or in the open air. 9. That in institutions in which the inmates are brought much into association the epidemic has more quickly attained its height, has prevailed more extensively, and been sooner over than in those in which the inmates are more secluded from one another.' The converse is also true, for Dr. Parsons states that those living away from the possibility of infection and in small groups, such as lighthouse keepers and deep-sea fishermen, enjoyed a remarkable immunity from the epidemic. It must be remembered that what is true of influenza is also true

[1] See p. 101 of the *Report on the Influenza*.

of other air-borne infections. It is known that whooping-cough and measles are prevalent and fatal very much in proportion to the density of population on a given area, and it has lately been shown that, although the mortality in so-called model dwellings is below that of the working-class dwellings as a whole, the occurrence in them of infantile infectious diseases, including diphtheria, is distinctly above the average. This is not to be wondered at, for with a large number of persons under a common roof the risks of infection must be proportionate to the number of inhabitants.

The modern city is distinctly a modern invention; it had no counterpart in ancient times. The ancients crowded behind walls from fear of their enemies. The modern city has no walls and no need of them, and there does not seem to be the same necessity for over-crowding that there was in the ancient city. As a matter of fact, however, the modern city both here and in America exhibits a degree of overcrowding on space such, as I believe, has never existed in the history of the world. The main cause of the difference between the ancient and modern city is the possession *of water under pressure*. In ancient Rome an excellent supply of water flowed from the neighbouring hills along open aqueducts supported on arches, and it is probable that the greatest head of water to be found in ancient Rome was represented by the height of the aqueducts above the level of the ground. Most of the buildings in ancient Rome were what we should consider low. The dwelling-houses were rarely of more than one storey, or at most two storeys, high, and it was only in public buildings— temples, baths, basilicas—that an attempt was made to carry the edifice to any great height. In mediæval cities, three-storeyed buildings represented the average maximum, for without water under pressure very high buildings were impracticable owing to the labour of carrying water to the upper storeys and the difficulty of removing refuse. Formerly a house was scarcely habitable unless it had a courtyard of some kind. This was essential for sanitary purposes —for a well and for the bestowal of refuse pending its removal by the scavenger; most of the decent houses in mediæval London had a considerable amount of curtilage, as may be seen by consulting the maps of Aggas, Ryther, Newcourt, and others. The steam-engine and the cast-iron water pipe have together created a new order of things. The steam-engine gives us any head of water we desire, and the modern iron fittings are able to stand any pressure to which we may find it necessary to subject them. There is no longer any difficulty in taking water to the top storey of any house, however lofty; no longer any difficulty in washing the filth out of any dwelling-house, however big; there is no longer any absolute necessity of limiting the height of buildings; no longer need of any curtilage to a house, nor for any outlet other than a pipe. This, indeed, is magical. 'My estate,' says one, 'has risen in value from 100*l.* an acre to 5*l.* a foot.' 'My water shares,' says another, 'pay me 12 per cent.' 'My mansions in the sky,' says the builder, 'have sold for double what they cost me.' All the gas shareholders and those who have invested in electric supply companies join in the chorus of delight. This is, indeed, a civilising and wonderful age. Let us build a temple, and place in it a steam-engine, an iron water pipe, and an hydraulic lift. Under these very money-making conditions it is not to be wondered at that the craze of the present time is for tenement houses, which are called model dwellings, flats, or mansions, according to the class for which they are intended.

Whatever they may be called they are all of the same type, and they are calculated to trouble the soul of the sanitarian who knows the danger of overcrowding on space and under a common roof.

Let us hear what Edinburgh can teach us on this subject. In a very excellent paper on the Hygiene and Demography of Edinburgh, which was communicated to the seventh (London) Congress, and which is now published in the twelfth volume of the Transactions of the Congress, is (p. 66) the following passage: 'Edinburgh enjoyed for many years the unenviable notoriety of being subject to periodical outbursts of fever. These assumed in all cases the epidemic form, and entailed a large mortality among the citizens. . . . There can be no doubt that one main cause of these repeated outbreaks was the manner of housing the inhabitants, which, copied from the French, consisted in piling tenement above tenement until a large overcrowded population was confined in a limited space, and could only communicate with the outer world by a narrow stair. . . . When infectious disease of any kind broke out in such circumstances, it spread with great rapidity, and quickly assumed the epidemic form.' Tenement houses are common in Paris, where the overcrowding on space as a consequence is on an average twice as great as in London, and where the death-rate is considerably higher than in London. (In 1881 I find the death-rate was 28 per 1,000 in Paris.) The condition of New York has already been alluded to. It is doubtless gratifying to national vanity to have a capital city composed of huge edifices which may be made to look palatial from the outside, but there is also no doubt that the comparatively low death-rate which London has hitherto enjoyed is to be attributed to the fact that as a rule each family has had a house of its own.

It is strange that the craze for high buildings has reached its maximum in America, where cities have sprung into sudden existence on the boundless prairie, where space is practically unlimited. The main causes of this seem to be hurry, avarice, an insane spirit of competition, and the difficulties connected with domestic service in a country where the citizens are more willing to slave for a machine than serve a mistress. Life in hotels and boarding-houses is more common in America than here, and co-operative housekeeping has reached a high stage of development. Water, warmth (by steam coils), gas, electricity, telephones, and hydraulic power are supplied to groups of houses from common centres. This is a convenient arrangement, but one which makes for the overcrowding of houses on space. Domestic architecture is in danger of becoming a lost art in cities where the inhabitants are content to be stored in human warehouses erected by engineers and as full of machinery as a factory. In the most modern city in the world (Chicago), the builders of which have had the opportunity of avoiding all the errors of the past, it is curious to note that the engineer-architects have exceeded all previous efforts in the matter of high buildings and disgraceful overcrowding. The 'Masonic Temple' in Chicago has twenty storeys and is 265 feet high, and is merely the latest and worst of many similar buildings in the city. In such a city the sewage difficulty is necessarily very great, but we learn that a special canal is to be constructed for its reception, and this is to discharge into the upper waters of the Mississippi! It seems astounding that such an outrage on their noblest river should be tolerated by the American people. The volume of the sewage may be as nothing when compared with the volume of the river into which it is to be discharged, but one would ask, Can this ingenious race, who like to be considered in the van of progress, do nothing better with the sewage of a great city than this? And are we to conclude that their hygiene is founded more on expediency and money getting than on any true scientific, economic, or moral principles? As regards the state of public health in Chicago, the following figures may help one to form an estimate of it. The annexed table, compiled from the official mortality returns, shows that for 1890 the death-rate was 18·2, and for 1891 22·2. It has not been possible to make any correction for age distribution, but it seems highly probable that Chicago is largely peopled by vigorous adult immigrants who crowd into it in search of employment. The zymotic death-rate was 4·09 in 1890 and 5·61 in 1891, the latter figure being not far from double the zymotic death-rate of London in 1890. The death-rate from typhoid fever was 840 per 1,000,000 in 1890 and 1,596 per 1,000,000 in 1891, the latter figure being nearly five times as great as the largest typhoid fever death-rate ever recorded in London. Croup and diphtheria last year killed 1,087 per 1,000,000 in Chicago, which is a figure enormously in advance of any that have ever been recorded in London. These figures seem to show that no amount of ingenious engineering can counteract the evils of overcrowding on space.

Death-rate per 1,000,000 *Living.*

	All causes	Zymotic	Diphtheria and croup	Typhoid fever	Tubercular affections	Respiratory affections	Violence
Chicago, 1890	18,214	4,090	1,051	840	1,859	3,136	1,069
,, 1891	22,204	5,610	1,087	1,596	1,967	4,089	1,170
London, 1890	20,978	2,896	330	144	2,943	4,910	719

The following account of Chicago, taken from a paper on the Columbian Exposition read before the Society of Arts in December last by Mr. James Dredge, seems to show that the modern American business city is admirably calculated to rob existence of its pleasures: 'No description, in fact, would do justice to this city of 1,200,000 inhabitants, almost every one of whom appears endowed with preternatural activity; which owns a street eighteen miles in length, almost a dead level for the whole distance, and on which are houses twenty storeys high; whose traffic is as noisy as it is ceaseless, both on the smooth-running rope-worked railways and the ill-paved, jolting roads; where the roar of the locomotive and the scream of the lake steamer emphasise the fact that repose and silence are unknown, even in the dead of night; where a pall of smoke, the outpouring of a thousand factories and of ten thousand dwellings, reminds the Englishman of home. Picture all these things, and you can form some idea of Chicago, which has been raised in sixty years by the indomitable energy of Americans to the rank of the sixth city of the world in point of population. . . . Great as Chicago is, the period of her true greatness has yet to come. Its commencement will dawn when her inhabitants give themselves leisure to realise that the object of life is not that of incessant struggle; that the race is not always to the swift, but rather to those who understand the luxury and advantage of repose as well as of sustained effort. Real greatness does not depend on length of streets, nor height of houses, nor even on colossal fortunes; but rather on the wise application and equally wise conservation of energy and intellect. When Chicago ceases to be the city of perpetual haste, and adopts the pace which will be inevitably set for her by time, the names of her great workers will not be erased so early from the book of life, but will be preserved to give their beloved city many more years of really useful work. At present, I think there are no old men in Chicago, because they have no time to grow old; and giving themselves insufficient time for leisure, they have, as a necessary consequence, little opportunity for the higher culture which is born of leisure.'

It needs to be pointed out that the overcrowding of our cities, which is admittedly the greatest of all sanitary evils, is the direct result of sanitary legislation. When, some forty years since, it was recognised that water-carried sewage inevitably poisoned the wells, no attempt was made to protect the wells from pollution; but the surface wells were compulsorily closed, the inhabitants were sold in bondage to the water companies, and the houses were compelled to pollute our rivers. These measures have given a fatal facility for overcrowding, and wherever a line of sewers is taken the speculative builder follows; for, not being hampered by questions of water-supply or filth-disposal (the sanitary authorities relieving him of all responsibilities in this matter), he is able to erect buildings without any curtilage, and of any height almost that he chooses. Be it observed that this state of things is not limited to London or the big towns. The notion that houses are as well without curtilage as with it has infected the country, as anybody may see for himself when travelling by railway. The train runs through miles of country, with scarcely a house to be seen; but when we stop at the developing district, growing up round the country station, we find the houses packed almost as closely as they are in the centres of towns. As an instance of this, a village might be mentioned where land can be bought at very little above its agricultural value. This village has been selected by a railway company for the establishment of 'works,' and the company is erecting rows and rows of artisans' cottages with at most a few square feet of curtilage each. The onus of draining these cottages will ultimately be thrown on the local authority, with the additional onus, in a few years' time, of providing the inhabitants with allotments ; and all the expense of this will fall upon the ratepayers as a whole. Clearly, the best place for an allotment is round the dwelling, and some compulsion should have been put upon this wealthy company to provide them. A couple of hundred acres of land (enough for 1,000 cottages) might, I suspect, have been purchased in this particular spot for less than 10,000*l.*, and would have added 10*l.* to the cost of each cottage. What is this to a rich company, especially if it be the means of increasing the health and contentment of their workpeople ? Would it not be money well spent, and spent to the ultimate advantage of the shareholders? We have engendered a stingy habit of mind towards the question of space round houses, which almost amounts to a national insanity.

A country town which the writer knows well, where 200 acres of land with a house and farm buildings recently sold for 4,000*l.*, has sanctioned the erection by a building society of 200 dwellings on six acres of land. This building estate will, if successful, have a population of 1,000 persons, or nearly 170 to the acre, a degree of overcrowding about three times as great as that of the average of London. The building society will make a good profit, while the sanitary authority which sanctioned this scheme will ultimately have to deal with the infectious diseases of this crowded area, and will not only have to provide, at the expense of the ratepayers as a whole, for the safe bestowal of the filth from this area, but will also, possibly, be compelled to provide the tenants with allotments.

Can anything be worse from the hygienic point of view than the modern tenement house or monster hotel, where each floor ventilates into the floor above and the floor below, while the 'lifts' effectually drive the air from one floor to another without renewing it? In such places, where hundreds live under a common roof, fresh air is impossible. Dr. Ogle, in his paper on 'The Relation of Occupation to Disease and Mortality,' read at the late Congress, pointed out that the mortality of hotel servants was higher than that of any other class, being nearly four times as great as that of the clergy, who enjoy the lowest degree of mortality. This high mortality is not to be wondered at, for the hotel servant never breathes fresh air, and lives day and night in an overcrowded dwelling redolent of humanity, dinners, gas, tobacco, and drains, one or all, most commonly all. Alcoholism may have something to do with this high mortality, but we must all admit that the unwholesome conditions of life are probably the main cause of the alcoholism. Drinking is said to be on the increase in this country, and it is probable that it would be found to bear a very definite ratio to the overcrowding on space. The advantages of hotel life are many, and the charms of living in a pleasant community are very great. These charms have always been recognised, but when we contrast the 'college,' as seen at Oxford or Cambridge, with its ample area and the opportunities for society or seclusion at will, with the modern hotel, raised tier above tier, in which tranquillity and fresh air are alike unknown, it must be confessed that, æsthetically and hygienically, the mediæval architect was ahead of his successors. These remarks apply not merely to those great hostelries in cities, where persons are content to jostle each other during a short paroxysm of business or pleasure, but also to establishments erected in health resorts and by the seaside, which have been designed for the reception of delicate persons and invalids, and which are well calculated to bring chronic maladies to a termination.

Let us look at the question of overcrowding on space from an economic point of view. It is calculated that to raise the body vertically requires twenty times the force necessary for walking on the level, so that whether we mount stairs to the height of 100 feet or walk 2,000 feet along the level the labour is the same. But I shall be told there are 'lifts,' and, that being the case, it becomes quicker and easier to move vertically than horizontally. It costs money to be carried, however, and a 'lift' is a most extravagant machine, because it does not accommodate the force expended to the weight to be raised. In a report recently issued by the South London Electric Railway, it is stated that the cost of working the lifts is 10 per cent. of the gross receipts! We are becoming alive to the fact that the fogs of London are getting more frequent and more virulent, in spite of the steady increase in the use of gas as a fuel. This is due to the overcrowding of houses and the multiplication of chimneys on a given area. In the country, and in most towns, the smoke is diluted to a point which practically gives us no trouble. It need hardly be said that a house of nine storeys will give off three times as much smoke as one of three storeys having the same area. In the centre of London we have been building enormous piles of offices and tenements, adding storeys to old houses, and putting buildings upon every garden and back yard, while at the same time the area of the city has enormously increased. What right have we to grumble at the increased density of the fogs ? We have deliberately caused it, and neither Royal Commission, parliamentary committee, nor anthracite coal will put it right. Even though we got rid of the 'blacks,' we should still have the irritating invisible products of combustion—by far the most harmful of the elements of a fog—to deal with. We have made our bed, and we must lie upon it. The damage done to property by fogs, the extra cost of washing and painting entailed by living in the dirtiest capital city in the world, the serious loss of trade, the cost of using artificial light in the daytime, as well as the injury to health, must be all reckoned as among the penalties we pay for overcrowding. 'Cleanliness is next to godliness,' says the old adage, and we are bound to admit that the filthiness of the air of London is a not unimportant factor in causing the moral degradation in which not a few of the inhabitants are sunk.

One main object of overcrowding houses is the saving of time in the transaction of business, but there are limits which cannot be overstepped without defeating our object. The check put upon freedom of locomotion by overcrowding, and the time occupied in going from point to point, more than counterbalance the shortness of the distance which we have to travel, to say nothing of the fact, previously alluded to, that vertical movement requires twenty times the force of horizontal movement. In London rapid locomotion is getting daily more and more impossible, and there are so many vehicles and persons in the streets that we are all in the way of each other. Our streets are always more or less blocked by building operations, which must be in proportion to the cubic contents of the buildings on a given area. Then, again, we have beneath our streets sewer pipes, water pipes, gas pipes, hydraulic power pipes, and pneumatic tubes, as well as wires for telegraphs, telephones, and electric lights. The necessity for interfering with these various subterranean arrangements in any one street must be largely proportionate to the number of dwellers whose front doors open into the street. These being almost unlimited, it follows that our streets, instead of being only occasionally 'up,' as the phrase goes, are only occasionally 'down.' Again, the number of vehicles which stop at any given door is proportionate to the number of persons who live behind it. A set of 'mansions' with 200 inhabitants may have scarcely more frontage than a modest house accommodating ten persons, but the vehicles stopping at them will be twenty times as many. It is the halting vehicles rather than the moving ones which offer the greatest obstacles to traffic. In civilised London we have very few back doors; our coals are shovelled in and our garbage hauled out under the dining-room windows; and these tedious operations, performed in the main thoroughfare, are in themselves great obstacles to traffic. The wear and tear of the streets is in proportion to the traffic, and in some of the leading thoroughfares the traffic has become so great that it is wonderful that any material can be found hard enough to stand it. These considerations make it clear that the increasing difficulties of locomotion in London are due to the overcrowding of houses on a given area.

From the æsthetic point of view the effects of overcrowding are disastrous. Proportion is one of the most important elements in architectural beauty, and few will be found to admire the Gargantuan architecture of the engineer-builder,

In spite of a somewhat dull uniformity, Regent Street still remains the finest street in London in virtue of its outline and proportions and the possibility of sunlight getting access to it. Contrast Regent Street with Northumberland Avenue, that drafty, gloomy gorge of bricks and masonry, in which the only really pretty building is the cabmen's shelter at the end. It is mere waste to erect fine buildings in situations where they cannot be seen without effort. Every good building ought to have a proper 'setting.' How would Westminster Abbey look without its greenery and turf? Can any building look really well without some verdure or floral decoration? Can any building that has them look really ill? In London our buildings are getting too high for their confined situations; they block out the sunlight, their many chimneys foul the air with soot, and, as a consequence, our palaces are grimy, and gardening, the most beautiful and the most health-giving of all arts, has become impossible. Wandering in the streets of London, and looking at the dim outline of our grimy public buildings looming through the mist, one is reminded of the lines from the opening of 'Macbeth':—

> Fair is foul, and foul is fair:
> Hover through the fog and filthy air.

When the air of a city gets so foul as to hinder and arrest vegetation, we are hygienically in a 'parlous' state, as Touchstone would have said. Not only do the green leaves of plants absorb carbonic acid from, and give off oxygen to, the air, but their roots are no less useful in draining and purifying the soil upon which we live. We have heard a good deal about the dangers of an impure soil, and the invariable advice with regard to it is to put something which is impermeable, such as concrete or asphalte, to prevent the rising of possible impurities from the earth. This is analogous to putting dead bodies into impermeable coffins, whereby the danger of their decomposition is not prevented, but merely delayed. The only way to purify the soil is by cultivation, aeration, and the growth of plants. I believe that an evergreen creeper, such as ivy, does more to keep the foundations and walls of a house dry and pure than do any of the patent impermeable applications.

As London, in spite of all its 'betterments,' is getting steadily less habitable year by year, there is very properly a growing tendency to sleep out of town and journey to and fro to business. Not more than 37,000 people sleep in that square mile which we call 'the City;' but as more than 300,000 find permanent daily occupation there, it is obvious that there are some 260,000 daily workers who have to be transported to and from the suburbs. If we reckon cost and loss of time in transit, it will not be an extravagant estimate to suppose that this journeying involves a loss of one shilling per head per diem, or, collectively, 13,000*l*. a day, 78,000*l*. a week, and over 4,000,000*l*. per annum. When a man lives out of town he becomes a slave to a railway company, he has to be punctual himself in order not to miss trains which are seldom punctual, and he finds out what it is to bear 'the whips and scorns of time.' As he wastes his time in being shuttled to and fro he must be queerly constituted if he does not wish for the good old times when it was possible to 'live over the shop.' Our modern facilities of communication cause us to be always in a hurry, and it is certainly a suggestive fact that having invested nearly 1,000,000,000*l*. in railways, on which we go shrieking to and fro; having so perfected the penny post that it has become a veritable nuisance with its endless delivery of documents, which are mainly worthless waste paper in the form of advertisements; having annihilated time and space by the telegraph and telephone; and having abolished darkness by means of gas and electricity—there should arise an ominous cry for a limitation of the hours of labour and a demand to do by Act of Parliament that which, 'in the good old days,' the sun did for us.

There is yet another consideration which shows the cost of overcrowding houses. When growth occurs in an organised body, such as a tree or an animal, all parts increase *pari passu*, and there is no necessary loss of health or vigour. But in a city growth is more like an inflammatory swelling, the cells of the part increase, while there is no adequate increase of the channels which bring nutriment. Stasis is apt to occur, and unless relief be given by surgical means the death of the part will follow. The modern city is ever in need of surgical interference; new thoroughfares have to be cut in order to give relief to congested districts, and these operations can only be carried out by means of an expenditure of money which is simply appalling in amount. The attempt to improve the sanitary state of a city by a process of 'Haussmannising' must be futile. If we make a fine boulevard through the slums the inhabitants are merely piled in heaps on either side; the width of the street is increased by increasing the overcrowding of inhabitants under a common roof, which is the greatest of all sanitary evils. Not long since a scheme was foreshadowed in one of the magazines for turning the Euston Road into a boulevard by appropriating all the front gardens and back yards, and erecting huge tenement houses on either side. No proposition could be worse from a sanitary point of view. It cannot be too earnestly insisted upon that the air of the grimiest back yard is better than that of any room, even though its (generally closed) windows command a view of a stately boulevard. The breathing of fresh air, so important to all, is doubly important to very young children. Children under four are too young to play in the streets; their mother's home duties are too arduous to allow her to take them to the park, and the little girl who might do so has been driven to school by the law. The result is that the little children have to breathe the fetid air of a small living room from week's end to week's end, instead of being able to play in a little yard, where the mother could watch them through the open door, as formerly was the case. The value of an open space is largely in proportion to its proximity to the dwelling. A little back yard or garden of one's own is worth infinitely more than Hyde Park a mile off, and the old graveyard converted into a garden a hundred yards down the street is far more precious to the poor town dweller than Epping Forest or Burnham Beeches.

This mention of disused graveyards leads one to say a few words on the disposal of the dead. Cremation is just now being strongly advocated, but on insufficient grounds, and for the following reasons :—1. Rational earth burial has never been shown to be productive of any evil, and the facts of bacteriology point to the conclusion that it is not a source of danger to the living. Even the scandalously irrational burial which has been permitted in the past has caused very little definite mischief, not even in those cases where the act of consecration has not been able to keep off the dangerous invasion of the graveyard by the railway contractor. 2. Cremation is wasteful, and involves a needless expenditure of fuel, and would help to still further pollute the air of our cities by products of combustion, always poisonous to breathe, and not always odourless. Dead bodies should be left to nitrify in the earth and provide nourishment for trees which, while affording us timber and firewood, would at the same time effectually cleanse the soil of its impurities and render it fit to be used for burial a second time. The advocates of cremation say that this economic argument is false because the products of combustion are indestructible, and must ultimately go to nourish plants somewhere and somehow. This is doubtless true, and by parity of reasoning we might do well to throw our money into the street instead of paying it into the bank, and we might advocate the fertilising of the market gardens round London by setting fire to the manure in the London mews. 3. The necessity for burying the dead which hitherto has existed has had the indirect effect of separating the living and preventing to some extent the overcrowding of houses. If our cemeteries are to be replaced by furnaces with tall chimneys pouring noxious gases into the air, and if the ground which would be occupied by cemeteries is to be covered by houses, it is clear that we shall gain nothing from a sanitary point by the introduction of cremation. 4. Although the writer is a professor of medical jurisprudence, he thinks the objection that cremation will enable the poisoner to escape detection is a minor objection, though real.

The overcrowding of houses means loss of liberty, for the closer we crowd together the more likely we are to suffer from the sanitary negligence of others, and the more necessary does it become to regulate our actions by by-laws and Acts of Parliament. I need hardly insist on the fact, so ably enforced by Herbert Spencer, that personal liberty is necessary for social evolution, and that Acts of Parliament are evils only to be tolerated for the avoidance of greater evils. The growth of Acts of Parliament of late years, and the new offences which they constitute, was recently forcibly pointed out by the chief constable of Liverpool in a letter to the *Times*, in which he stated that in 1890 there were as many as 11,279 cases in Liverpool under the Education and Health Acts alone. Overcrowding has necessitated the compulsory notification of disease and the appointment of a crowd of inspectors, and there are those who wish to compel us to take out a license for our houses, which is only to be given when they are fitted with the particular sanitary apparatus which happen to be in fashion. It is well to bear in mind that the overcrowding of which I am complaining is largely the result of compulsory Acts of Parliament which curtailed our liberty in the matter of water-supply and sewerage, while they gave to the speculative builder opportunities by which he has not failed to profit. When a couple of years since the influenza made its reappearance among us, a distinguished engineer wrote to the papers with the kindly object of comforting us by the assurance that the perfection of our sanitation would certainly lessen the ravages of the epidemic. It need hardly be said that the author's opinion was directly opposed to his, for influenza being an air-borne disease, and the overcrowding of houses and the fashion of many families living under a common roof having largely developed since its last appearance, it seemed probable that the epidemic would be more severe than formerly, and more difficult to eradicate.

Can nothing be done to check the overcrowding of houses and persons? Although we have passed many laws for the improvement of the public health, we have not sought as yet to put any very severe restraints on the builder and landowner, and we have not sought, as would seem but reasonable, to limit the number of inhabitants who may be accommodated on a given area of land. It is the object of all traders, whether they be public companies or individuals, to have the largest number of customers on the smallest possible area, and therefore any attempt at restriction cannot be popular. Neither would such restriction be popular with the sanitary authorities, who have to collect rates; for rateable value is unfortunately directly proportionate to the overcrowding of houses on a given area. We have borrowed large sums of money, and the rates have more than kept pace with our sanitary progress (the rates of the author's own house have risen more than 30 per cent. in the last twenty years), and as houses pay rates the authorities encourage building by every means in their power; they foster overcrowding and perhaps call it 'betterment.' Sanitary authorities are now being encouraged to embark in business as purveyors of water, gas, electricity, and locomotion; and as their success in business will depend upon the number of customers they can crowd into their respective areas, they are likely to be more biassed than ever in favour of overcrowding.

One of the functions of a sanitary authority is to protect the public from the rapacity of public companies; and one would fear that the desire to make money out of the needs of the inhabitants must hamper the judgment and impartiality of the authority. This fashion of sanitary authorities to become purveyors must be regarded as an interesting socialistic experiment, but not one which is likely to conduce to the public health or public convenience. We have many sanitary laws, and the present does not seem the time to advocate a return to the liberty of the past. Unless the sanitary condition of our towns is to get steadily worse, we must have a law to prevent overcrowding. Such a law seems to me to be the necessary complement of all the others. It seems needful to insist that every building shall have a minimum curtilage in proportion to its cubic contents, and it is evident that any such rule, to be of value, must be of universal application throughout the country. Such a rule, we shall be told, is impossible because of the high price of building land in cities; but to this objection the answer is that the exorbitant price of building land is the result of the absence of any such rule. The value of building land is, *cæteris paribus*, proportionate to the number of inhabitants it will accommodate; and if the accommodation of a given area is to be unlimited, the price of land will be almost unlimited also. It is evident that any rule for making the minimum curtilage proportionate to the cubic contents must not be too exacting, and must be very simple and capable of universal application. Suppose we were to say that the curtilage of a building shall in no case be less than an area expressed by a figure representing 1 per cent. of the cubic contents of the building. This is certainly a very modest demand. Let us see how it would work. By such a rule as this a building like the Masonic Hall at Chicago, which is 114ft. × 117ft. × 265ft., and which (to take round numbers for simplicity) contains about 3,500,000 cubic feet, would require 35,000 square feet of curtilage, and since the area of the building itself is 13,000 square feet, the size of the entire plot of ground which it would require would be 48,000 square feet, or about an acre and one-eighth. Assuming that this building can accommodate 1,000 persons (probably a low estimate), it is evident that, even with the compulsory curtilage which I suggest, we should still have a condition of overcrowding on space which must be regarded as indefensible.

The average London dwelling-house, which has about twenty feet frontage with a height of sixty feet and a depth of fifty feet, contains 60,000 cubic feet, and ought, according to the rule I have suggested, to have a minimum curtilage of 600 feet—*i.e.* a back yard as wide as the house, and thirty feet long. Such a house would accommodate about ten persons, and would occupy 1,600 square feet of ground, or, with its necessary proportion of street, 2,200 square feet, or, as near as may be, the twentieth of an acre. Even under these circumstances we should have 200 inhabitants to the acre, which is a number certainly largely in excess of what ought to be allowed. It is evident that, unless we know the maximum number of inhabitants which is to be permitted on a given area, it is impossible to make any reliable calculations as to water-supply and sewage disposal. It does not seem to me more unreasonable to set a limit to overcrowding houses on a given area than it is to require a definite cubic space for the inhabitants of common lodging-houses. This rule could not, of course, be applied to existing houses, but the cubic contents of such houses or of houses built to replace them ought not to be suffered to be increased unless this rule or something like it be observed. If we allow the cubic contents of buildings to be indefinitely increased, regardless of the question of curtilage, as is being done in every part of London, it is certain that the condition of public health must steadily deteriorate. To imagine otherwise would be to give the lie to those elementary facts of physiology and hygiene which are universally admitted. To purchase plots of vacant ground at exorbitant rates, and miscall them 'lungs,' is no true remedy for overcrowding. The only true remedy is to have the open ground dovetailing with the houses. Let us also remember that overcrowding in the daytime is no less harmful than overcrowding at night. The citizens of London have seldom had an eye to anything except immediate profit. They allowed Moorfields and the Drapers' garden to be sold for enormous sums, without a thought for the health of the thousands of young men and women who daily toil in that overcrowded spot called 'the City'; and they have taken no heed of the Registrar-General's returns, which are telling them that the death-rate at the prime of life (*i.e.* between the ages of thirty-five and sixty-five), which has been steadily rising for the whole kingdom, is nearly 70 per cent. higher in London than it is in the best country districts.

No regulation other than the insistence upon a proportion being observed between cubic contents and curtilage would serve to check overcrowding. An arbitrary rule to limit the height of buildings is not sufficient, because if the builder be not allowed to soar he will commence to burrow, as he is doing in Chicago, and as he has done for many years past in London. A crying need in London is for back doors and back streets between the houses for delivering goods, and beneath which could be laid the various subterranean pipes and wires, the repairing of which is so frequent a cause of blocking the main thoroughfares. A system of back streets would do more to facilitate the traffic of the thoroughfares than a similar addition made to the width of the main streets.

It will be urged that these views are impracticable, because overcrowding is necessary for business, and it is indeed asserted by some that the commercial prosperity of this country and of America is largely due to the fact that the business men are content to literally crawl over each other in an ecstasy of frenzied competition which is making existence a burden. It is certain, however, that this question of overcrowding on space is one which must be dealt with in the future in no niggardly spirit, and that our national and commercial prosperity very largely depend on our willingness and ability to face it. The inadequate housing of the working classes has probably more to do with their chronic discontent than is generally supposed. A real 'home,' be it remembered, is a very different thing from a room in a barrack. A home that is clean and bright and beautiful, the beauties of which have increased year by year under the owner's fostering care, where the children have space to grow up healthily, and where there is some escape from the dirt and din of the workshop—a home in which it is possible to take a pride, and which is a source of daily pleasure— is something which a man will weigh carefully in the balance before he lightly determines on throwing up his employment. A decent home without sufficient space is impossible, and the sooner that great industrial companies and employers of labour grasp this idea the better. That the labouring classes are beginning to grasp it is shown by the recent agitation in the East-end of London against replacing the two-storeyed houses by industrial barracks. The fashion of piling people in heaps, and offering them libraries and bagatelle-tables as a compensation, must come to an end, and one looks forward hopefully to the working of the Allotment Act, especially when it is recognised that the proper place for the allotment is round the house. Thirty perches (rather less than a fifth of an acre) is the maximum amount of garden which a man can cultivate at his leisure, and it is the little bit of garden which converts a dwelling into a home, which provides a delightful occupation for a man's leisure hours, which freshens the air round the dwelling, which keeps a man from the tavern and the sixpenny betting hell, and which is capable of turning all the refuse of the house into profitable garden produce. Not much improvement is to be looked for until the conscience of the educated and upper classes has been roused. As long as persons in high position are content to leave their broad acres and swelter for the season in highly decorated 'bijou residences,' where the servants are poked away underground or in miserable attics, not much improvement can be expected. Many of the persons who come to London for the season are philanthropists whose hearts 'bleed' for the miseries of the East-end. These people forget that when they come to town they make it necessary for humbler people who minister to their wants to come to London also; and that, if they who are wealthy are content to live on a very insufficient area, their poor dependents must necessarily be packed together in great discomfort and at the cost of their physical and moral well-being. Let those who are sorry for the state of overcrowding in cities stop out of them, and spend their incomes in the country. If they can succeed in making those who live round them in the country happy and contented, and thus check the rush for the towns, they will do infinitely more good than by subscribing to city missions or 'People's Palaces.' Unless we can succeed in checking the overcrowding of cities by some such regulation as that which has been suggested, it is probable that all the money we have spent and are spending on so-called sanitary improvements is money thrown away, and that, in so far as modern hygienic methods have enabled people to live closer packed than heretofore, they have worked mischief rather than good. Those who have the control of developing districts will do well to keep them as rural as possible, and encourage private gardens by every means in their power. The house ought to be taxed in proportion to its size, and not the garden round it; for a garden is a great boon in a town, even to those who do not own it, because it serves to keep the dwelling-houses asunder. A person who owns a garden may, if he be so minded, be quite independent of the sanitary authority; and a wise sanitary authority will do well, by a remission of taxation, to encourage him by every means in its power. So long as sanitary authorities are content merely to encourage the jerry builder by clearing up his messes and making good his deficiencies, not much improvement is to be expected. Such policy may be 'good for trade,' but it is not good for the public health.

Finally, one hopes that it may be increasingly possible for a man to own his dwelling and the ground around it, and that the transfer of land may be much facilitated. Without wishing to embark on a discussion as to the relative advantages of large and small holdings, it seems necessary to point out that no one is so likely to do justice to the land and to improve it as the absolute owner, whether he owns a thousand acres or only one. If a man owns a bit of land, be it farm or garden, he is bound in self-defence to make the best of it, and he can have no object in quarrelling with, cheating, or shooting his landlord. This last is a point which political economists not unfrequently neglect. We are often told that large holdings are more economical than small ones, and it is pointed out that on the farms of our large proprietors the yield per acre is greater than the yield on the land of the peasant proprietors of France or elsewhere. Our big farmers, however, have not of late years grown rich, and it is doubtful if free trade is the sole cause of this. Is it not possible to farm too high for profit? We may drive a steamship ten knots an hour at a profit, but lose by attempting to get fifteen knots out of her. So I fancy that much of the profit of farming is frittered away in a lavish expenditure on artificial fertilisers, in the desire to get excessive crops. Small holdings and gardens and what has been called the dovetailing of the houses and the land, have two advantages which are often lost sight of. The first is, that the producer and consumer are largely identical, so that the middleman and free trade are alike powerless to do him harm. The second is, that all the refuse of the house is available for the land, and in proportion to its amount is the expense of buying 'artificials' and the risk of keeping stock done away with. The key to good agriculture is thrift, and no arrangement more thrifty than that which has been indicated can be conceived. The key to good sanitation is agriculture, and therefore no effort should be spared to maintain a rural element even in our great industrial centres.

CHAPTER III

ON THE SHORTCOMINGS OF SOME MODERN SANITARY METHODS

THE chief aim of sanitarians has ever been, and ever will be, the securing for the masses of the people the two chief necessities of life—pure air to breathe, pure water to drink. Whether or not we are able to secure these two necessities depends very largely upon the method which we adopt for the treatment of putrescible refuse; and it is to this point, and the modern fashion of mixing putrescible refuse with water, that the present chapter will be devoted.

It may be well to remind the reader that all dead organic matter is putrescible, and that, when putrescible matter is spoken of, all organic matter, inclusive of excrement, is meant.

Nature moves in a circle, animals feed on each other and on vegetables, vegetables feed on the dead bodies of animals and vegetables, and on the solid and gaseous excrements of animals. Animal and vegetable life are complementary, and mutually support each other. This is a law of nature, and in making this assertion I run no risk whatever of being contradicted.

The laws of nature are inexorable; *i.e.* they are not to be set aside by human prayers—not even by that best of all prayers, labour. Those who disobey the laws of nature, or who enter into a contest with her, are sure to be worsted in the end. If we fight with nature we court calamity.

Those who fight with nature may be compared to Sisyphus, who, according to the old mythology, was condemned in the lower world to a never-ending contest with the force of gravity—

> With many a weary sigh and many a groan
> Up the high hill he heaves a huge round stone;
> The huge round stone resulting with a bound,
> Thunders impetuous down, and smokes along the ground.

By means of great expenditure of time and money, we may wage for a period with nature a war which may be apparently successful. The war can never be really successful, it will never terminate, nature in the end will assert her eternal sway, and crushing defeat must be our lot.

As the inevitable destiny of putrescible matter is to become the food of vegetables, a destiny which we can delay at the most only for a brief period, our proper course in dealing with it is clearly not to attempt to prevent or even to delay the inevitable. Such a course is to disobey the laws of nature, to fight with her and court ultimate defeat. Our wiser plan is to help nature in her work, and thus win her smiles.

It has been the wise custom in all ages of the world to dispose of putrescible matter by burial in the earth. Dead bodies have in all ages been buried, and the greatest of all lawgivers and sanitarians, Moses, gave most explicit directions that excremental matters should be treated in the same way.

This is a not unimportant fact, and although we do not in this country follow the whole of the Mosaic law, nevertheless, that law is so pregnant with marvellous wisdom, that we ought not to discard any item of it without first questioning ourselves most strictly as to our reasonableness in so doing. The latest advances of modern science seem to show that in this particular Moses was absolutely in the right.

It has been shown, I think conclusively, that the decomposition of organic matter, whether in the earth, air, or water, is brought about by minute fungoid organisms, the growth of which has the effect of resolving the highly complex organic compounds into soluble salts or gaseous bodies, which can be absorbed by the roots of plants.

Now when putrescible matter is buried in the earth it undergoes decomposition without the occurrence of putrefaction—that process which is at once offensive to the senses and dangerous to health. This is effected by means of fungi, which produce oxidation of the organic bodies. If sufficient air has access to the pores of the soil, and if sufficient moisture be present, the nitrogen takes oxygen to form nitric acid, and this, combining with the bases, forms soluble nitrates. The carbon also in a similar way forms carbonic acid and carbonates.

A good account of these active organisms which are ever present in the soil will be found in a paper by Professor Wollny,[1] of Munich. These organisms are so incalculably numerous that their activity must be exceedingly widespread. Koch found enormous quantities, even in winter, in the soil not only of crowded places like Berlin, but in that also of remote fields. At the observatory of Mont Souris 750,000 were found in a gram of earth, and at Genevilliers from 850,000 to 900,000.

[1] 'Ueber die Thätigkeit niederer Organismen im Boden,' *Deutsche Vierteljahrsschrift für öffentliche Gesundheitspflege*, vol. xv. p. 705, 1883.

If the action of the microbes be checked by antiseptics, the vapour of chloroform or heat (100° C.), the chemical changes in the earth cease.

That the formation of nitrates and carbonic acid from organic matter in earth to which air has access is due to microbes has been proved by direct experiment. When, however, organic matter is mixed with earth, and air is admitted in insufficient quantity or entirely excluded, the decomposition is of another kind; and besides small quantities of carbonic acid and carburetted hydrogen, there is formed water, ammonia, free nitrogen, and a great quantity of a black carbonaceous peat-like matter (the so-called sour humus).

Schlösing found that the nitric acid in the soil disappeared when the air was replaced by nitrogen.

The kind of organism seems to vary with circumstances. As long as air is freely admitted, the mould-fungi (*Schimmelpilze*) preponderate; and when air is excluded, the schizomycetes (*spaltpilze*) increase.

The formation of nitric acid in organic earth mixtures depends on the amount of oxygen which is present in the air admitted. Thus Schlösing found by experiment that the formation of nitric acid varies as under:—

Oxygen	1·5%	6	11	16	21
Nitric acid	45·7 m.g.	95·7	132·5	246·6	162·6

The nitrification which took place with a limited supply of oxygen was due probably to the air already mixed with the earth before the experiment began.

Miller and Boussingault have shown that no nitrification takes place in thoroughly soaked earth to which little air has access, and that when oxygen is absent the nitrates in the earth are reduced. The formation of carbonic acid also depends upon the admission of air (containing free oxygen), but some carbonic acid is formed even though all air be excluded.

Oxygen in air	Pure N.	6%	11	18	21
Carbonic acid	9·3 m.g.	15·9	16	16·6	16

Nitrification is assisted by a moderate amount of moisture. It attains its maximum when the moisture reaches 33 per cent., and above and below this the process of nitrification and formation of carbonic acid is hindered.

Temperature has a great influence on oxidation in the earth. Oxidation reaches a maximum with a temperature of about 50° C. (120° F.), and stops at 55°.

Oxidation goes on most quickly in the dark.

Thus, oxidation depends not only on the presence of the organisms, but also on the presence of other factors, such as suitable aeration, suitable moisture, suitable temperature.

These factors may all be suitable, or some may suit and others may not suit the oxidation process.

The decomposition of organic matter in the soil is governed by that factor which is at its minimum.

The process of decomposition is much influenced by the physical condition of the soil, as, *e.g.*—

(a) Permeability for air and water.
(b) Nature and permeability of subsoil.
(c) Slope.
(d) Aspect.
(e) Warmth dependent on aspect, mineral composition, colour and moisture and nature of the crop. Barren soils are warm, while those covered with green crops are cool.

That the variations of the ground water have a bearing on the oxidation processes cannot be doubted, when we reflect that the soaking of the upper layers of the earth is much influenced by the height of the ground water. When all the layers of earth are soaked, putrefactive processes, through the medium of schizomycetes, take place. When the ground water falls, and the air again enters the pores of the soil, the growth of those organisms is favoured which assist in the oxidation of the soil.

All changes which organic matter undergoes in the earth are thus seen to be brought about, almost exclusively, by the life of organisms, the activity of which is ruled by the same natural laws which govern the growth of higher plants. There can be no better illustration of the true economy of nature than this action of the microbes in the soil on the conversion of organic matter into soluble salts and gases which serve as food for plants.

The growth of the microbes depends upon the concurrence of those conditions which, by experience, we all know to be favourable to the growth of higher plants. There must be a good supply of free oxygen, sufficient, but not too much moisture, and a summer temperature. In well-tilled ground, broken up so as to admit air to its pores, and in a 'fine growing season,' in which sunshine alternates with showers, this process of oxidation is at its maximum. The microbes are active beneath the surface manufacturing plant food from organic matter, and the favourable conditions above soil and below cause a vigorous growth of crops.

When, on the other hand, the weather is unfavourable, and when in consequence of excessive cold, excessive drought, or excessive wet, crops are not developed as they should be, the microbial life is also checked, and the change of the organic matter is delayed, and it is stored up for future use in more favourable seasons. This is the explanation apparently of the fact well known to farmers, that the effect of organic manures is more permanent than that of the so-called artificial manures, which at present are so much in vogue. The organic manure remains entangled in the soil, and is not readily washed out of it in winter when the temperature is low, or even in unpropitious summers. It cannot be washed out until microbial growth has changed it into soluble salts, and when this change takes place, which it does in 'good' weather, the roots of the growing plants seize hold of the ever-forming soluble salts and appropriate them to their own use. In fact, the farmer who uses organic manures from the farmyard or elsewhere need trouble himself very little with agricultural chemistry or experiment.

He may feel certain that if he buries his organic manure *directly it is produced* it will not be wasted. It will not give off ammonia to the air, nor will the juices be washed away by rain to the same extent as when it is left above ground to be a nuisance. There seems to be no doubt whatever that all heaps of manurial matter which give off ammonia and other gases to poison the air, and perhaps do more serious mischief, are allowing valuable matter to escape, which ought to be undergoing oxidation in the earth. There can be no doubt whatever that to the agriculturist stink means waste, and it is to be hoped that, when the bucolic mind has imbibed this great and important truth, the country will be more evenly pleasant than it is.

The reason why farmers allow putrescible matter to fester in heaps appears to be—

1. That the matter has to wait until land is clear and circumstances permit of its being dragged to the fields;

and (2) that when the matter is thoroughly rotten and most offensive, a *more rapid and visible* result is produced, notwithstanding that the total result is probably less than if it had been applied to the ground at once. It is certain that putrescible matter intended for manure must waste more above ground than when buried immediately beneath it. Rich farmers are now building sheds over their yards to prevent the access of rain to the manure, and are providing tanks for the reception of liquid which drains away. This involves a very great expense, and it is at least doubtful whether the result is better than that got by the immediate application of such matters to the soil—a process which involves no extra expenditure of any kind—a most important matter, because the only acceptable test of good husbandry is the balance-sheet.

Mr. Warington, F.R.S., in his valuable little book on 'The Chemistry of the Farm,' says: 'The most complete return to the land would be accomplished by manuring it with the excrements of the men and animals consuming the crops' (p. 28); and again, 'Farmyard manure is a "general" manure; that is, it supplies all the essential elements of plant food. . . . The effect of farmyard manure is spread over a considerable number of years, its nitrogen being chiefly present not as ammonia, but in the form of carbonaceous compounds, which decompose but slowly in the soil.'

The immediate return is often less than when artificial manure, consisting of soluble nitrates and phosphates, is used, but the important point seems to be that the return is tolerably sure to come in the long run.

The late Professor Voelcker, in the article 'Manure' in the 'Encyclopædia Britannica,' gives an interesting table of the experiments of Sir John Lawes and Dr.

Gilbert, spreading over a period of twenty-four years, in which is shown the effect of different manures on crops. The most successful results with artificial manure were got by applying nearly 1,400 lb. weight per acre of mixed ammonia salts, superphosphate and sulphates (potash, soda, and magnesia). With this manure there was an average production of $37\frac{1}{2}$ bushels of wheat, weighing on an average 59 lb. per bushel, and multiplying these two figures together we may say that the production of wheat averaged 2,212·5 lb. The production of barley averaged $41\frac{1}{2}$ bushels, weighing $53\frac{3}{8}$ lb., and multiplying these figures we may say that the average production was 2,588 lb. Where the land was manured with 14 tons of farmyard manure the average production of wheat was $35\frac{1}{4}$ bushels, weighing 60 lb., giving a figure of 2,115 lb., and of barley, $48\frac{3}{4}$ bushels, weighing $54\frac{3}{8}$ lb., giving a figure of 2,650 lb.

This farmyard manure, when used for wheat-growing, gave a yield of 97 lb. less than when the best artificial manure was used ; and when used for barley-growing it gave 62 lb. more than when artificial manure was used. These figures are certainly not such as should discourage us in the use of farmyard manure, especially when we remember that the average agriculturist is not likely to apply his artificial manures with the knowledge and judgment of Messrs. Lawes and Gilbert ; and that in the use of farmyard manure it is not easy for him to go very wrong. Again, farmyard manure is stuff which *must* be used, while chemicals are things which *must* be bought, and need to be analysed when bought.

Among the 'Memoranda' (issued in June 1891) of results obtained by Sir John Lawes at Rothamstead will be found the following, which still further bears out the point in question. Barley and wheat have both been grown continuously upon the same land for thirty-eight years, upon a series of experimental plots, each plot being manured with different mixtures of chemical manures, with farmyard manure, or (in the case of one plot in each) left unmanured continuously. The results are given in tabular form, in which is set forth the average yield of each plot for the whole thirty-eight years and for each half of the thirty-eight years (nineteen years 1852–70, and nineteen years 1871–89).

There are in all (of barley and wheat) fifty-five plots of which these particulars are given. The most important fact seems to be that on fifty-four of these plots the average yield, whether of barley or wheat, was *less* in the second nineteen years than in the first nineteen years, showing that the fertility of the land had in every instance *except one* deteriorated. This solitary exception is the plot upon which barley has been grown continuously, and which has been manured year after year with fourteen tons of farmyard manure to the acre. Upon this plot the yield for the first nineteen years averaged 48 bushels of dressed grain and $27\frac{3}{4}$ cwts. of straw per acre, while for the second nineteen years the average was $49\frac{1}{4}$ bushels of dressed grain and $29\frac{3}{4}$ cwts. of straw.

The most successful of the artificially manured barley plots was No. 4, A.A.S., which had been dressed continuously with 275 lb. nitrate of soda, 400 lb. silicate of soda, 200 lb. sulphate of potash, 100 lb. sulphate of soda, 100 lb. sulphate of magnesia, and $3\frac{1}{5}$ cwts. of superphosphate. Upon this plot the yield for the first nineteen years averaged $50\frac{1}{4}$ bushels of dressed grain and $30\frac{1}{2}$ cwts. of straw, while for the second nineteen years the average was $44\frac{4}{5}$ bushels of grain and $28\frac{1}{4}$ cwts. of straw. Upon most of the other plots the yield was far inferior to the above, and upon the plot which had been unmanured continuously the average yield for the first nineteen years was 20⅛ bushels of grain and 11⅞ cwts. of straw, and for the second nineteen years 13¼ bushels of grain and 7 cwts. of straw.

On the plots devoted to wheat the average yield in the second nineteen years has, without exception, been less than in the first nineteen years, and the plot manured with farmyard manure does not show that pre-eminence of fertility which is the case with the barley.

Thus, on the most productive of the artificially manured wheat plots, the yield in the first nineteen years averaged 39 bushels of grain and 41⅝ cwts. of straw per acre ; and in the second nineteen years the average was 34 bushels of grain and 38½ cwts. of straw. This plot has been manured each year with 200 lb. sulphate of potash, 100 lb. sulphate of soda, 100 lb. sulphate of magnesia, 3½ cwts. of superphosphate, and 600 lb. ammonia salts.

On the plot dressed continuously with farmyard manure the yield of the first nineteen years averaged 35¾ bushels of grain and 33½ cwts. of straw, and in the second nineteen years 32½ bushels of grain and 30¼ cwts. of straw.

The object of the above extracts is to show the high value of organic manure—which they undoubtedly do. In the case of barley there can be no doubt on this point, but in the case of wheat the results with organic manure are surpassed by some of the artificially manured crops.

In the season of 1890, the results of which are tabulated in these 'Memoranda,' the results obtained with organic manure were so remarkable that they deserve to be quoted, although they have not the value of the nineteen-year averages.

The yield of barley on the plot dressed with farmyard manure was 53 bushels of grain and 29⅛ cwts. of straw (against 47½ bushels and 22¾ cwts. on the best of the artificially manured plots), and the yield of wheat on the plot dressed with farmyard manure was 50 bushels of grain and 48⅝ cwts. of straw against 48 bushels and 46¼ cwts. on the best of the artificially manured plots. The season of 1890 seems to have been characterised by a rather dry spring and a rather wet summer.

It is a great mistake to suppose that farming is in any way comparable to a chemical experiment. In experiments conducted in the laboratory the chemist is able to control *all* the conditions of the experiment, but in farming the condition which above all others influences the result, viz. the weather, cannot be controlled.

When chemical manures are used with judgment and applied at the right moment, and when the weather is favourable, there is no doubt that the result is often surprising and gratifying. When, however, the weather is unfavourable, when the drought is so great that the chemicals cannot be dissolved, or when the rain is so heavy that they are washed out of the soil, the result is not encouraging. If organic manures are used, they waste but little in bad seasons, and much remains in the ground for next year's crop. The farmer, however, who applies chemicals in a bad season gets neither crop nor residuum of manure for next year. Mr. Warington says that 'farmers have a prejudice in favour of the latter' (*i.e.* organic) 'manures, but it is clear that the quickest return for capital invested is afforded by the former class' (*i.e.* inorganic).

Surely we have no right to blame the farmers for their prejudice, which seems to be in all respects reasonable. The doctrine has obtained in this country of late years that it is good economy to waste all our home-grown organic manure, and to import chemicals from

South America for the purposes of agriculture. This is a strange doctrine; but as most of our farmers are now too near bankruptcy to pursue this course, one may hope that ere long they will begin to clamour for that which we now waste so wickedly.

One more word before bringing these remarks on farming to a close, remarks for which no apology is needed, because their bearing on the subject of sanitation must be obvious.

It will be noted that in the hands of Lawes and Gilbert farmyard manure gave better results with barley than with wheat. May not the fact that farm animals are largely fed with barley-meal have something to do with this? There are experiments which show that minimal ingredients in manures are not without effects which are often surprising. There are *a priori* grounds for thinking that the best manure for barley must be the excrement of a barley-eating animal, for in that excrement must be all that is necessary for barley. It is to be regretted that some agriculturist does not make the experiment of growing wheat with the excrement of a wheat-eating or bread-eating animal. As a gardener the author has grown potatoes with the excrement of a potato-eating animal, and certainly the result has been most encouraging.

One has been obliged to draw illustrations as to the practical result of burying organic matter from the agricultural employment of farmyard manure, because facts based upon exact experiments with the organic refuse of our towns is not forthcoming.

The point to be insisted upon is this: that the proper destiny of organic refuse is immediate burial just below the surface of the soil.

Most of the shortcomings of modern sanitary methods

are due to the fact that in our dealing with organic refuse we commit a scientific error—*i.e.* we pursue a course which is in opposition to natural law.

This error consists in mixing organic refuse with water.

When organic refuse is mixed with water it undergoes changes which differ widely from the changes which it undergoes when mixed with earth.

According to Wollny, whose paper has been quoted previously, the process of oxidation of organic matter and the formation of nitrate take place most readily when a moderate amount of moisture is present. The most favourable amount is about 33 per cent., and if the moisture rise above or sink below this amount, the process of nitrification and the formation of carbonic acid are hindered. When water is in excess the amount of free oxygen is insufficient to favour the growth of the necessary fungi, and, in place of oxidation, putrefaction takes place, with the formation of ammonia, free nitrogen, carbonic acid, and carburetted hydrogen. Under these unfavourable circumstances it is possible that the nitrates which may have been formed may be again reduced.

This process of deoxidation takes place in mixtures of putrescible matter with water, and takes place also, it is said, in soil which is thoroughly soaked with sewage (*i.e.* putrescible matter mixed with water). In the face of these facts, it is not to be wondered at that 'sewage farming,' which is farming under acknowledged difficulties, has not proved a commercial success. We must, indeed, be in doubt whether, when the circumstances are more than usually unfavourable, it exercises any very great purifying action upon the putrescible mixture. In the treatment of putrescible refuse, so that it shall not be a danger or annoyance, what we have to aim at is nitrification rather than putrefaction, and it is certain that by mixing with water putrefaction is encouraged and nitrification delayed.

It certainly seems to be almost incontestable that the proper course to pursue with regard to organic refuse—putrescible matter—is the very reverse of that which we do pursue. We clearly ought to encourage oxidation, and make putrefaction impossible.

Putrefaction is certainly a great cause of ill health. It was the putrefaction of wounds (now happily almost unknown) which converted our hospitals into something little better than charnel-houses. It is the putrefaction of organic refuse mixed with water in cesspools and sewers that causes that long list of ailments which we ascribe to the inhalation of 'sewer air.'

The opinion is held by many that the dejecta of typhoid patients and cholera patients do not become dangerous to others until putrefaction has set in, and such an acute observer as was the late Dr. Murchison held the opinion that common putrefactive changes taking place in dejecta were a sufficient cause of typhoid independently of the admixture of any specific poison.

The putrefaction of organic refuse, when mixed with water, has, there is reason to think, been the chief cause of the development of modern sanitary 'progress.' Our forefathers were not given to this method of treating putrescible matter. House-slops trickled along open gutters, and excremental matters were deposited in dry pits. At the beginning of this century the water-closet came into use.

Mr. W. Haywood, quoted by Dr. Farr, says: 'Water-closets were invented about 1813, and became general in the better class of houses about 1828–33. The custom at first obtained of building cesspools having overflow drains put below their doming, by which means the solid matters were retained and the supernatant liquid only ran off.

'In the year 1849, what may be said to be an organic change in the system took place. In 1848 the City Commission of Sewers obtained its Act for sanitary purposes, which became operative on January 1, 1849, and then for the first time was discharge into the sewers legalised. Previously a penalty might have been enforced for such a usage of them, but henceforth, within the City of London, those incurred a penalty who failed, upon notice, to construct the drainage of premises in such a manner as not to discharge all waste waters *and fæcal matters directly into the public sewers* [*i.e.* directly into the sources of water-supply], of which the full utility was therefore for the first time recognised by statute. This Act was speedily followed by others for the remaining area of the metropolis and for the entire country.'

'It will be noticed,' says Dr. Farr, 'that the deaths from cholera and diarrhœa increased in London in 1842, increased still more in 1846, when the potato crop was blighted, and in 1849 culminated in the epidemic of cholera.'

Dr. Farr says, further, 'a system of sewerage is the necessary complement of a water-supply.'

'Almost coincidently with the first appearance of epidemic cholera, and with the striking increase of diarrhœa in England, was the introduction into general use of the water-closet system, which had the advantage of carrying night soil out of the houses, but the incidental and not necessary disadvantage of discharging it into the rivers from which the water-supply was drawn.'

Mortality per 1,000 from diarrhœa in London (Dr. Farr):—

Year	Rate	Year	Rate
1838	·215	1853	1·011
1839	·201	1854	1·257
1840	·238	1855	·804
1841	·238	1856	·806
1842	·353	1857	1·181
1843	·410	1858	·759
1844	·340	1859	1·211
1845	·397	1860	·496
1846	·997	1861	·928
1847	·898	1862	·607
1848	·853	1863	·821
1849	1·705	1864	·981
1850	·813	1865	1·206
1851	1·085	1866	1·306
1852	·983	1871–80—Dr. Ogle	·940

Thus in the decade 1871–80, 33,168 persons died of diarrhœa in London, the death-rate from this cause being ·94.

If the death-rate of 1838 (·215) had obtained in the decade 1871–80, the deaths from this cause would have numbered only 7,600, and there would have been a saving of 25,568 lives.

Since the introduction of the water-closet, and probably, as a direct consequence of it, we have had four severe epidemics of cholera, a disease not previously known, and enteric or typhoid fever, previously almost or quite unrecognised, has risen to the place of first importance among fevers in this country.

The evils which have arisen from cesspools and sewers has caused an enormous amount of attention to be devoted to what are known as 'sanitary appliances,' sewer constructions, &c., and so great and so well recognised are the evils of sewers that many of our friends are anxious that we should be compelled, by Act of Parliament, to protect ourselves from the mischief which previous Acts of Parliament have produced.

Not only does the putrefaction of organic refuse tend to fill the air of our houses and towns with foulness, but this mixture of organic matter with water is attended with other bad consequences.

This arises from the fact that much of the organic matter which we mix with water is distinctly poisonous. The zymotic theory of disease has of late years assumed more definite shape, so that we may now leave what was called the zymotic *theory* and consider the actual facts.

There is no doubt that the actual infective elements of many zymotic maladies consist of microbes, fungoid bodies belonging to the class of fungi known as Schizomycetes, that class which grows in organic mixtures.

These microbes are infinitely small; millions of them may live in a cubic inch of putrefying liquid. Under favourable circumstances they will live for long periods. They will not only live but multiply, and it is at least a question, and a grave one, to what extent these infective germs undergo an increase when mixed with organic liquids such as sewage or milk.

The fact that the zymotic poisons are *particulate and alive* is one which has most important bearings on the subject under discussion. If the poison were a chemical poison, then dilution would practically do away with its power for harm. No amount of dilution is capable of destroying a zymotic poison; in fact it is not impossible that the mere mixing of organic refuse which contains a zymotic poison with water may be the means of keeping it alive and possibly causing it to multiply.

When a mass of organic matter, charged with zymotic particles, is mixed with water and washed out of a house, the water will carry the poison with it where-ever it may chance to flow or trickle, to watercourse, well, or any other source of drinking water; in fact, the dissemination is as perfectly and thoroughly done as if dissemination of poison were the main object which we had in view.

When dealing with organic matter impregnated with zymotic poisons, mere dilution with water increases rather than diminishes the danger.

As long as the poisonous organic refuse is concentrated, its repellent qualities are such that there is little chance of its gaining access to the human body. The microbes contained in it are theoretically capable of infecting an almost indefinite quantity of water, and this large quantity of water masks the repellent qualities of the stuff, and thus the danger of infection is greatly increased.

This dissemination of poison by water is one of which we have had very bitter experience in this country.

There is little room for doubt that, in this country at least, water has been the great carrier and disseminator of the poison of cholera.

In 1849 the mortality in London was highest in those districts getting their water-supply from the Thames between Battersea and Waterloo Bridge.

In 1853-54 the same phenomenon was observed. In 1866 the chief mortality was in the district supplied with water taken from the River Lea. With regard to this latter epidemic, we are in possession of many details, and the following is a summary of the facts as given by the late Dr. William Farr in his report on the cholera epidemic of 1866 :—

' Several cases of cholera and choleraic diarrhœa had occurred over London in May; and on June 27, at 12, Priory Street, Bromley, one poor Hedges, a labourer, and his wife, both of the age of 46 years, died of "Cholera Asiatica," the former after fifteen, the latter after twelve hours' illness. These cases are minutely described by Mr. Radcliffe, who traces the discharges into a water-closet of 12, Priory Street, and thence 300 yards down the sewer into the Lea (a tidal river which ebbs and flows) at Bow Bridge, half a mile below the Old Ford reservoirs. He attaches great importance to these first cases, and they undoubtedly sufficed to pour into the sewers and waters millions of zymotic molecules, which day by day grew more and more frequent in the Lea, by every hour's choleraic discharges on both sides of the river.' A few days later water was supplied to the district from a reservoir the bottom of which was pervious to the waters of the Lea, and then resulted an outbreak of cholera and diarrhœa which caused the death of over 4,000 persons.

It is not necessary to give further instances of the dissemination of disease by water-carried sewage, sanitary literature is full of them.

What is true of cholera is also true of typhoid, and one need only say in reference to this subject that (if we accept, as we are bound to do, the statements put forward with regard to the cholera epidemic of 1866), if the excreta of the Hedges' family had been buried or burnt, the waters of the Lea would not have been infected, and possibly 4,000 lives would have been saved.

The first principle in dealing with epidemic disease is that which is expressed in the words *principiis obsta*— resist the beginnings. The object of this is evident, and is well expressed by Shakespeare in the words—

    A little fire is quickly trodden out,
    Which, being suffered, rivers cannot quench.

The mixing of bacteria with water may be looked upon certainly not as a resistance of the beginnings, but rather as a nursing and favouring of them, which, 'being suffered,' most surely 'rivers cannot quench.'

The great principle of *principiis obsta* has been most rigidly observed by surgeons in dealing with those forms of blood-poisoning which arise in connection with wounds, and which were known as 'hospital diseases.' To Lister belongs the credit of recognising that the great thing to be aimed at was the checking of putrefactive changes in the discharges from the wound, an end which has been attained by adopting what are known as antiseptic precautions in the treatment and dressing of wounds. A foul wound is looked upon as a great source of danger to the patient himself, and formerly the poisons generated in the wound of one patient were carried by sponges and instruments (which, be it remembered, were '*clean*,' as far as any indications appreciable by our unaided senses were concerned) to the wounds of others; and thus it followed that the mortality from what was wrongly spoken of as 'hospitalism' was enormous. Now, however, putrefaction in wounds is practically at an end, owing to the use of antiseptics and to an improved appreciation of what cleanliness really means; and, as a result of this, hospitalism has disappeared.

How marvellous have been the results which have followed on the adoption of the principle of preventing putrefaction in wounds is well shown in a table given in 'Erichsen's Surgery.' This table is taken from a statistical work by Max Schede on amputations, and shows conclusively what are the advantages of antiseptic precautions. The statement has been somewhat simplified for the sake of those who are not acquainted with medicine.

Uncomplicated Cases of Amputation.

Cause of Death	Old treatment, 377 cases	Antiseptic treatment, 321 cases
Blood poisoning	105	3

Thus it appears that the mortality from blood poisoning under the old treatment was 28 per cent., while, under antiseptic precautions, it is less than 1 per cent.

Antiseptic measures are used in other than purely surgical cases, and to take one instance only, Dr. John Williams has informed the author that since their introduction into the General Lying-in Hospital the deaths from that terrible disease, 'puerperal fever,' have practically ceased.

This great result has been brought about by attention to the leading principle of *principiis obsta*.

In former days the treatment most in vogue for wounds was 'pure' water; but now it is recognised that water is pre-eminently the encourager, and sometimes the main cause, of putrefaction, which of all things the surgeon tries to avoid.

The foulness of our rivers is largely due to the mixing of putrescible matter with water, *i.e.*, to water-carried sewage, and there can be no doubt that, as water-carried sewage increases, the difficulty of obtaining pure water increases also. Water-carried sewage so fouled the Thames 'between the bridges,' that, after the bitter experiences of 1854, the in-take of the water companies was moved to a point above the tideway. Since then the population all along the Thames Valley has enormously increased, and if we who get our drinking water from the Thames escape disease, it can only be regarded as due to a happy accident, and not to the observance of any fixed principle to effectually prevent the fouling of the river. The precious liquid with which the author is supplied from the Thames costs nearly ten shillings per thousand gallons, and he is very careful to have every drop which is used for drinking purposes both boiled and filtered.

It is not possible to have much faith in the various modes of 'treatment' which sewage undergoes in those establishments which local boards love to erect for this expensive amusement.

The addition of chemicals, if in sufficient quantity to destroy living organisms, must make the water still more unpotable than before, and can only be of use by making the liquid so utterly nauseous that to drink it would be impossible.

Mere filtration unaccompanied by oxidation cannot be regarded as any safeguard after the experience of the Lausen typhoid epidemic, in which the poison of the fever filtered through a mile of earth, which was sufficient to check the passage of particles of wheat flour. Wide irrigation over a large area of land, as is practised in 'sewage farming,' is probably the best method of treating sewage, but this cannot be regarded as absolutely safe under all conditions for reasons previously indicated.

If antiseptics have been previously added to the sewage, this must increase the difficulties of 'farming' with it, as, if the antiseptics have been added in sufficient quantity to destroy disease organisms, this would effectually check the growth of those other organisms upon which the fertility of the soil depends.

It is more than doubtful whether there is any absolute safety in obtaining water from deep wells. The

Dudlow Lane well, near Liverpool, having a total depth of 448 feet, was fouled by percolation from cesspools, and percolation from a defective sewer would certainly prove equally disastrous. Surface wells are now not regarded as at all safe, but our suspicions with regard to them were not aroused until after the introduction of the plan of mixing water with putrescible matter. There was no soakage from an old-fashioned dry pit; there must be soakage from a cesspool or 'dead well.'

The only way of securing pure water is to make quite sure that there is no fouling of water-sources. If this were done, then pure water would be at once plentiful and cheap. It is now very dear, and is getting scarcer every day.

Dr. William Farr said, 'a system of sewerage is the necessary complement of a water-supply.' One would rather feel inclined to say that an extraordinary water-supply is the necessary complement of water-carried sewage, because with it our ordinary supplies quickly get fouled. In London we have effectually fouled all our wells, and the state of the Thames is such that a man must be in the very extremities of thirst or else insane before he would drink from the Thames anywhere between Teddington Lock and Gravesend. The state of our noble river is a deep reproach to us, and must remind us day by day of the serious blunders we have committed. As long as it remains as it is, we certainly have no claim to be followed as an example in matters sanitary. London should rather serve as a warning, as did the drunken Helot to the Spartan Youth.

The fouling of our sources of water-supply has driven us far afield for water, and this, no doubt, has been a great cause of the lessening of our mortality of late years, but it would be unwise to talk of security because we have had no serious epidemic since 1866, an absurdly short period in the history of a nation. It must not be forgotten that pure water is as necessary for animals as it is for man, and that, if we persist in fouling our rivers, the poor farmer may have to pay a 'water-rate' for providing an artificial water-supply for his horses, cattle, sheep, and even poultry. Many diseases of animals are communicable to man, and it is daily becoming more evident that our health is very intimately bound up with the health of our animals, and that their sanitary condition is scarcely less important than our own.

From a financial point of view, water-carried sewage has not been encouraging. It has increased the rates, increased the cost of our houses, and put us to great expense for water. The 'treatment' of sewage before it is finally discharged into our rivers is everywhere an expense and nowhere a source of profit, and we find that public sewers which cost millions, cost thousands to keep them in repair.

The sewers we have built with borrowed capital. We have seized all the glory and patronage of disbursing enormous sums, and have left posterity to pay the bills. This is a doubtful policy, and a most immoral one, but it is little use to raise one's feeble voice against the custom which is now so much encouraged of hanging a debt round the neck of our successors. It may be defensible to raise a loan for building town halls, schools, and similar edifices, of which posterity will reap the benefit, but to raise loans for the purpose of wasting most valuable fertilising matter by means of works which will be a constant expense, and never a source of profit, is a very doubtful expedient.

It is to be hoped that the custom will soon obtain of

compelling each generation to bear the charge of its own sanitary experiments—and blunders.

Sewers are constant sources of impoverishment to the soil, and the soil, be it remembered, is the only *permanent* and reliable source of wealth in any country. The waste of valuable matter which takes place in London and our big towns must make us blush. One could wish that this waste were limited to our big towns, but it is not so. It is common throughout the country, even in rural districts. Free trade has made food very cheap indeed, and cheap food, especially *imported* food, ought absolutely to increase the fertility of a country, for obvious reasons which need not be particularised. The fertility of this country is not increasing, to judge by the agricultural distress. The farmers are crying out for 'protection.' The first kind of protection needed seems to be a protection from ourselves and from the sinful waste of fertilising matters which local boards, municipalities, and Imperial Parliament equally foster.

If we made a proper use of our organic refuse we should enrich posterity. As it is, we reap and we do not sow. If municipalities would bury organic refuse, and plant the seed of some forest tree suited to the soil and situation (which in these days of cheap foodstuffs would probably be the best branch of agriculture to pursue), they would earn the blessings instead of the curses of posterity, and they would beautify the face of nature instead of making it hideous with tall chimneys, pumping stations, and precipitating tanks. This piece of advice will, just now, fall very flat, for of all agricultural arts, forestry seems the deadest in this country.

As a defence for gigantic sewage schemes, it is often said that you can do nothing well without co-operation, and this is the excuse for compelling all, whether they want them or not, to contribute towards the cost of sewers.

If co-operation be for a good end, the result is a great good; but if co-operation be for a bad purpose, the result is a great evil. I need say no more.

The last charge which has to be brought against water-carried sewage is a serious one—viz. that it encourages overcrowding in cities, which is universally admitted to be the greatest of all sanitary evils, and one which cannot be counterbalanced.

Water-carried sewage encourages overcrowding because it enables us to build houses with no outlet except a hole for the sewage to run through. The growth of London must be a source of alarm to sanitarians, and it is impossible not to admit that our system of sewers has been a most important factor in its production. Look at Charing Cross, where a street of gigantic clubs and hotels has arisen, each without curtilage of any kind, and where a handsome profit has been made by setting the first law of sanitation at defiance. You will find the same thing to a greater or less extent throughout the Metropolitan area.

It is difficult to say why we are so prone to crowd into cities. In former days we crowded behind walls as a protection from our enemies. Those days are at an end, but the crowding is greater than ever. The common cant of the day is that in this nineteenth century we have annihilated time and space. Certainly in cities both are excessively precious. The telegraph, the telephone, and the steam-engine ought to have diminished overcrowding, but they have not. The stream is still mainly from the country towards the town, the attraction being the making of money and the spending of it.

It may be well to glance at the effect of this over-crowding in this city.

It is a common remark that London is a very healthy city, and as a proof of this assertion persons point to the death-rate, which certainly of late has not been excessive. The London of the Registrar-General, however, is a very extensive place, and many of the outlying parts are almost rural in character, so that, if you want to find the effect of living in a crowded city, it is not fair to take London as a whole.

The author is no believer in the healthiness of London. It is true that our death-rate has not been raised by any great epidemic of late years, but London is undoubtedly a city where an abundance of second-rate health exists. The crowds that throng the doors of hospitals increase, and in the medical profession there is a great outcry about 'hospital abuse,' which means, probably, that decent folk are not able to cope with the amount of chronic disorders with which they are beset. Again, the mobility of the population in the present day makes our vital statistics very uncertain. Many a healthy person is imported into London, and being wounded in the battle of life, returns to the country to die or recover, as the case may be. There is a scarcity of very young and very old people, and in order to appreciate the vital statistics of London, great allowances have to be made for the abnormal age distributions.

In order to judge of the effect of overcrowding, let us look at the vital statistics of the 'Strand' Registration District, which is about the centre of London, and from which one would have to walk very many miles to reach the country in any direction.

The 'Strand' enjoys many advantages. It is mainly a wealthy district, extending in irregular form from

Temple Bar to Buckingham Palace. It includes the whole of the Green Park and half St. James's Park. It has a gravel soil, and slopes gently, with exposure to the south, to the fringe of (potentially) the noblest river in the country. The worst and poorest parts are at the north-east corner.

The true death-rate of a London district is difficult to get. The Registrar-General, however, has been in the habit of publishing quarterly the 'true' death-rates of the London districts after complete distribution of deaths occurring in public institutions. The author has compiled a table by which one is able to compare 'The Strand' with the whole of London for nine quarters, and with Dorset (for ten years, 1871–80).

	Birth-rate	Death-rate	Zymotic death-rate	Deaths under one year to 1,000 births
London	32·5	19·9	2·7	151
'Strand,' and St. Martin's-in-the-Fields	23·7	21·8	2·6	192
Dorset (ten years, 1871–80)	29·53	17·46	1·68	108

The county of Dorset has been chosen for comparison because it is a 'healthy district,' and if we are to do any good we must always aim at a high standard. Again, the Dorsetshire labourer has always been a favourite stalking-horse for cockney politicians, and it may be well to show how much healthier he is than the Londoner, notwithstanding the rustic's supposed condition of chronic starvation.

This table is very interesting. Dr. Letheby said 'a high death-rate means a high birth-rate, and a high birth-rate is the invariable concomitant of prosperity.' This dictum does not evidently apply to the Strand.

Dr. Farr, on the other hand, pointed out that 'a low

birth-rate implies a small proportion of young adults and a large proportion of the aged.' This dictum again does not apply to the Strand, as we shall see by a reference to the next table, in which an endeavour has been made to make corrections for the abnormal age-distribution which obtains in that district, and which Dr. W. Ogle rightly insists is absolutely necessary before one can arrive at just conclusions.

The table speaks for itself.

'*The Strand*'—Mean Population 1871–80 = 37,461.

Ages	Actual numbers living at each age	Normal age distribution for a population of 37,500	Difference (+ & −) between actual and 'normal' numbers	Actual deaths in 10 years 1871–80	Deaths which would have happened if the distribution of ages had been normal	Death-rate at different ages	Death-rate of Dorset	Deaths which would have happened if the death-rate of Dorset had obtained in the Strand
Under 5	3,597	5,100	− 1,503	3,596	5,100	99·97	40·07	1,440
5–10	3,134	4,500	− 1,366	390	548	12·44	4·31	129
10–15	3,069	4,012	− 943	163	212	5·31	2·79	84
15–20	3,824	3,640	+ 190	317	299	8·29	4·43	167
20–25	4,426	3,337	+ 1,089	366	273	8·27	6·65	290
25–35	6,773	5,512	+ 1,261	963	770	14·22	7·50	510
35–45	5,121	4,237	+ 884	1,246	1,000	24·33	10·48	525
45–55	3,935	3,225	+ 710	1,338	1,088	34·00	13·04	520
55–65	2,311	2,212	+ 99	1,147	1,100	49·63	24·56	565
65–75	1,003	1,237	− 234	754	900	75·17	55·28	550
75	268	487	− 219	425	774	158·58	151·71	403
	37,461	37,500		10,705	12,074			5,203

From this table it appears that there was in the Strand during the decade 1871–80 a deficit of 3,812 children under fifteen, and of 453 of persons over sixty-five, while there was a surplus of 4,233 persons between fifteen and sixty-five.

This abnormal distribution ought, according to Dr. Farr, to give us a high birth-rate and a low death-rate. The very reverse is the case, and a critical examination of the figures seems to show that the death-rate in the Strand *is more than double what it is in Dorsetshire.*

It may be said that this high death-rate is due to the presence in the Strand of two hospitals (Charing Cross and King's College), and doubtless these have some material effect in producing the terrible adult mortality.

Hospitals, however, are generally placed where they are most needed, and I would point out that these institutions can hardly account for the enormous infant mortality; and certainly not for the deaths of infants under one year. Against the fact that the Strand contains two hospitals, is to be placed the not less important fact that it contains no workhouse. This institution is at Edmonton, where it helped to raise the death-rate from 15·8 to 16·9.

It need not surprise us that a population situated in the very centre of the vastest city the world has ever seen should have a high death-rate, and it may be well to look to the causes of death and again to compare the rates from different causes with those in Dorsetshire.

Death-rate from different Causes.

	Strand	Dorsetshire
Whooping-cough	0·62	0·29
Tabes	0·28	0·18
Phthisis	3·65	1·72
Hydrocephalus	0·61	0·22
Respiratory disease	5·92	3·15
Total of Tubercular and Respiratory disease	11·08	5·56
Small-pox	0·11	0·09
Measles	0·36	0·20
Scarlet fever	0·49	0·33
Enteric	0·38	0·19
Violence	1·61	0·49
Diarrhœa	0·92	0·35

No good would be got by extending this table. Suffice it to say that there is no single cause of death in the Registrar-General's tables which is not more active in the Strand than it is in Dorsetshire.

It is important to notice that the death-rate from whooping-cough and tubercular and respiratory diseases for the Strand is more than double that of Dorsetshire, a fact which is not to be wondered at in a population the bulk of whom only breathe pure air upon the rarest occasions, and who habitually breathe an air so foul that the sun often fails to penetrate it, and which is fatal to almost all flowers and a large proportion of trees.

One of the saddest indications of the dismal state of this overgrown city is the appeal, which is now so common in the newspapers, for funds to give poor London children *one* day in the country, with of course the not immaterial deduction of the hours spent in going and returning.

These tables may serve to dispel another popular fallacy, viz. that the sulphur-laden air of London has antiseptic powers, and helps to check zymotic disease.

As a fact, those zymotic diseases which presumably travel through the air (small-pox, whooping-cough, and measles) are particularly rife in London. The death-rate from these three causes was during 1871–80 :—

	In London	Dorsetshire
Small-pox	0·44	0·09
Measles	0·51	0·20
Whooping-cough	0·81	0·29
	1·76	0·58

In fact, the mortality caused probably by air-borne germs was exactly three times as great in London as in the healthy country district which has been chosen for comparison.

In this chapter we have attempted to show that the systematic admixture of putrescible matter with water is inadmissible—

1. Because it is antagonistic to a law of nature, encouraging putrefaction and delaying nitrification, and there can be no successful antagonism to nature.

2. Because the putrefaction set up in cesspools and sewers by mixing water with putrescible matter has been a direct cause of much disease.

3. Because the practice involves the most perfect dissemination of disease particles, and a neglect of the great principle, '*principiis obsta*.'

4. Because it is the great cause of the fouling of rivers and wells, and makes the obtaining of pure water increasingly difficult.

5. Because it is financially and economically disastrous, crippling the ratepayers and exhausting the land.

6. Because it is one of the chief causes of overcrowding, the greatest of all sanitary evils.

It may be asked, 'What useful purpose can be served by demonstrating these matters to Londoners? London is hopelessly committed to the principle of water-carried sewage, and must make the best of it.'

The obvious reply is that even London should not heedlessly increase her already insurmountable difficulties, and that happily the whole of England is not yet quite absorbed into London and other cities. There is a very general belief throughout the country that because London has adopted the system of water-carriage it must therefore be the best. This idea is unthinkingly adopted, and to its adoption the distinction of borrowing and disbursing a large amount of other people's money acts as a spur.

There has come within the author's own knowledge the case of a country town, in the midst of a poor agricultural district, which clamoured for a 'sewage scheme' for the purpose of polluting its sparkling water-course, where anglers pay large sums for the privilege of trout-fishing; its death-rate being at the time between 16 and 17.

In the Thames Valley, the region of villas and market-gardens, a whole crop of 'sewage schemes' have lately been put forward, notwithstanding that the more rational methods of sanitation would be easier and cheaper.

A few years since the author visited a lone farmhouse which a friend wished to take for the summer, and found that the proprietor, having taken the soil-pipe of a recently erected water-closet into a cesspool alongside a deep well sunk in the chalk, had rendered his house unlettable to any thinking person; and finally he would instance the case of another friend who took a moor in Scotland, and wished to have rational methods of sanitation, but the noble owner, bitten by the modern craze for water, would allow nothing but water-carriage, and accordingly laid his filthy pipes to foul the babbling highland burn, and deprive the soil of that of which it was in need.

Again, in institutions such as workhouses, barracks, schools, and the like, water-carriage is often adopted, notwithstanding the favourable conditions for rational methods. The ignorance of soldiers in this matter is an acknowledged cause of the sickness and mortality during campaigns.

There seems, in short, a very great necessity for

directing attention to the 'shortcomings' of water-carried sewage.

'What do you propose?' will be the next question. The answer is, 'Fair play and no compulsion.'

Great as is the good of spreading sanitary knowledge, the author has little faith in the efficacy and a potent belief in the dangers of sanitary legislation whereby blunders are stereotyped.

The first thing necessary is an equitable adjustment of sanitary rates.

Borrowing for the purpose of constructing sewers should be disallowed, and those who do not need the sewers should not be called upon to contribute towards them, at least not to the same extent as others.

The present inequitable adjustment of sewer rates is a premium on jerry-built houses without curtilage. Encourage the man who has a little bit of garden to make use of it.

Enforce the Pollution of Rivers Act against individuals, even against proprietors of highland moors.

Let us have a real inspection of nuisances and a harassing of evil-doers, and let us discourage by every means in our power the building of houses side by side, and almost back to back, with no outlet but a hole.

Let water be paid for by meter.

It is impossible not to sympathise with the agitations for giving allotments to the poorer classes. The best and most economical allotment is one close to the house, where refuse may be buried and in due time bring forth.

Those who advocate 'sewage farming' tell us that an acre is necessary for every 100 inhabitants. How infinitely better it would be if the 100 people could absolutely live on the acre of ground in (let us say) twenty cottages, each cottage having $\frac{1}{20}$ of an acre. How infinitely better for the man to till this little plot in his spare time than to occupy his leisure in less wholesome pursuits.

Let us calculate the produce of this plot of ground in terms of potato. An acre of a field will produce an average crop of 7 tons; the twentieth of an acre would produce 7 cwts., or 784 lb. As these would be for home consumption, and would save the man from disbursing money at a retail shop, we may take the value at the average retail price of 1*d*. per lb., or 3*l*. 5*s*. 4*d*., or for the sake of simplicity say 3*l*. To give $\frac{1}{20}$ of an acre to every five inhabitants would make a town inconveniently big, it may be said. But this would not be the case; 100 to an acre is 64,000 to a square mile, or making a very liberal reduction for space occupied by roads, let us say 50,000 to a square mile. This does not sound like an inconvenient scattering of houses. The inhabitants would make 30,000*l*. a year by the produce of the land, a gain of which Free Trade could not deprive them; and there would be no sewer rate, no plumbers' bills, and certainly a vast increase of health, happiness, and contentment.

CHAPTER IV

'THE LIVING EARTH'

SANITATION in large cities is, at the best, a makeshift, and no high level of health is attainable in a place where the chief object of hygienists seems to be to enable persons to live as densely packed as possible.

Those, therefore, who live in the country, and who enjoy the luxury of elbow room, should hesitate before they hastily copy the sanitary methods of the town, and heedlessly begin to foster overcrowding, the bane of all sanitary and social virtue.

This chapter is addressed to dwellers in the country, because the subject—the 'Living Earth'—is one which those who live on paving stones, tarred blocks, asphalte, or macadam, have to take upon trust.

The 'Living Earth'! Some may ask what is meant by this, and whether the epithet 'living' is applicable to the dark-coloured, inert mould which the countryman sees in the fields and gardens, and the town dweller finds in the flowerpot which holds his struggling geranium.

The reply is in the affirmative. We have arrived of late years at a certain knowledge of the fact that the mould which forms the upper stratum of the ground on which we live is teeming with life, and this fact is one of prime importance to sanitarians.

It has long been recognised by agriculturists that the

upper stratum of the soil differs from that immediately below it in fertility; and in treatises on gardening (notably in that admirable work written by William Cobbett nearly seventy years since) the warning is invariably given to be careful in trenching, not to bury the top spit of soil beneath the lower spit, because the top spit is by far the most fertile. The fertility in this case was supposed to be due to prolonged exposure to air, and the lower stratum of soil, if brought to the surface, would only become fertile after a considerable interval. It is interesting to observe that, although these early writers were unacquainted with the whole truth, they had grasped the most important fact, and their practice was sound. This is often the case, and it may be regarded as certain that we act rashly when we hastily abandon the custom of centuries, because some new fact dazzles us and distorts our vision.

In connection with William Cobbett, one may draw attention to a term which he uses more than once in the work referred to, viz. the *Fermentation* of the soil. Cobbett tells us that the earth begins to ferment in the spring, and that, before sowing, a thorough tilling and mixing of the upper layers of the soil are very necessary, with a view not only to the disintegration of the soil, but to a thorough leavening of the whole mass with fermentable matter. There is no doubt that this term 'Fermentation' as applied to the soil is perfectly apt, as we shall find further on.

The black vegetable mould which lies upon the surface of the earth is largely composed of organic matter, which is not to be wondered at, seeing that every organised thing, whether animal or vegetable, which inhabits this globe, falls, when dead, upon the earth, and becomes incorporated with it.

This black vegetable mould is largely composed of excrement, for not only is the excrement of the larger animals being constantly added to it, but this and the varied organic *débris* which compose it pass repeatedly, probably, through the bodies of animals which inhabit the earth, and especially of earthworms. Darwin, in his book on Vegetable Mould and Earthworms, has forcibly drawn attention to the enormous amount of work which worms perform in the aggregate. How they disintegrate the soil. How they riddle it with burrows, which admit air to the deeper recesses of the soil. How their castings, which are incessantly being thrown off, tend to level inequalities, and gradually to bury stones or whatever dead inorganic matter is incapable of solution, digestion, or disintegration. Earthworms are found almost everywhere, and they are probably the most important of the animals which live in the soil, but it need scarcely be said that there are many others, and everyone who has a garden must recognise the fact that gardening is only carried out at an enormous sacrifice of animal life, for with every thrust of the spade into rich garden mould a death-blow is dealt to many of its inhabitants.

The disintegration and aeration of the soil, which is effected by the quiet tillage of the earth-dwellers, is of the greatest importance to the agriculturist, for it is hardly conceivable that the delicate rootlets of plants could grow and extend unless the soil had been softened and pounded by the digestive fluids and the gizzards of the earthworms and their neighbours.

Seeing, therefore, that agricultural mould has all passed through the bodies of worms, and much of it through the bodies of other animals antecedently, we shall not be wrong in insisting that this so-called vegetable mould is mainly an animal excrement. The peculiar, sticky, glutinous quality of rich mould when moistened is probably in part due to this fact.

Although the amount of *Animal* life in the earth is considerable, it is as nothing compared with the richness of the soil in the lower forms of *Vegetable* life. The dead and excremental matter becomes the food of saprophytic fungi, which abound in the soil to a very great extent. This must be the case, for we know that saprophytes and their allies abound everywhere, and as the surface of the earth is the common reservoir of all forms of life, it follows that these low vegetable microbes must be more abundant in the earth than elsewhere, and more abundant at the surface than deeper down. In Watson Cheyne's editions of Flügge's work on micro-organisms (New Sydenham Society, 1890), this is very clearly stated: 'Enormous numbers of bacteria have always been found in the soil by the most various observers. Infusions made from manured field and garden earth, even though diluted 100 times, still contain thousands of bacteria in every drop, and the ordinary soil of streets and courts also shows the presence of large numbers. Bacilli are present in much the largest numbers; but in the most superficial layers, and in moist ground, there are also numerous forms of micro-cocci.'

These micro-organisms of the soil are very active in producing changes in organic matter added to the soil. These changes are usually in the direction of oxidation, occasionally the change is one of reduction. One thing is certain, that if the soil be sterilised by heat or other means, it is no longer capable of producing any chemical change in organic matter. This seems to be a fact of prime importance to the sanitarian. The oxidation and nitrification of organic matter in the soil is a biological question, pure and simple. It is an effect produced by the *living earth*; a process analogous to fermentation, which Cobbett seems to have appreciated.

Whether the nitrifying process which takes place in the soil is due to one or to many varieties of microbe is doubtful, but the latter supposition is probably correct, and experiments seem rather to point to the conclusion that, given favourable conditions—the free admission of air to a soil which is not unduly moistened—nitrification will go on. Many attempts have been made to isolate a nitrifying organism, and one of the latest, by Professor Percy Frankland and Grace Frankland, the results of which were communicated to the Royal Society in February 1890, appears to have been successful, for these observers isolated a 'bacillo-coccus,' the power of which in producing nitrification appears to be most remarkable. Whether this bacillo-coccus is one of many having similar power, or whether it stands alone, is not known, but in any case we must regard it for the present as the 'Nitrate King' among microbes.

It has been asserted that fungi of a higher class, mould fungi, are also active in producing the disintegration and oxidation of organic matter in the soil. It is possible, however, that the *Bacillus mycoïdes*, which forms threads closely resembling mycelium, has been mistaken for mould fungus. This bacillus mycoïdes is one of those which is constantly present, we are told, in garden soil.

It has been conclusively shown by Flügge, Koch, and others that the microbes are most abundant in the superficial layers of the soil, and that they tend to disappear in the deeper layers. They are practically absent in the deeper layers, unless the earth has been deeply stirred or trenched, or unless sewer or cesspool has conducted filth to the deeper layers without touching the superficial ones.

'Numerous filtration experiments on a large and small scale have shown most distinctly that a layer of earth $\frac{1}{2}$ to 1 metre in thickness is an excellent filter for bacteria, and hence the purification of fluids from bacteria must be still more complete in cultivated, and especially in clay soil, and where the fluid moves with extreme slowness. Further, it has been repeatedly shown that wells which are properly protected against contamination, from the surface and from the sides, furnish a water almost entirely free from bacteria; that further, wells of water containing bacteria become the purer the more water is pumped out, and the more ground water comes in from the deep layers of the soil.'

The vegetable living mould on the surface of the earth is in short a filter of the most perfect kind. It is very rich in saprophytic bacteria, whereas the subsoil at a depth varying from 3 to 6 feet is barren of bacteria, as well as of other kinds of life. The subsoil is mineral, inorganic, and dead; the mould upon the surface is organic, and teems with life.

It seems to be an undoubted fact, and one which in sanitary matters is fundamental and of the greatest practical importance, that, *from the point of view of microbial life, the first few inches of the soil is, so to say, worth all the rest.*

Anything which is thrown upon the surface of the ground soon disappears.

This is especially the case with water. The absorbing power of soil for water varies according to its mineral constitution. Loose sand and chalk absorb water very readily, and clay less readily, but the absorbing power of vegetable mould, or humus as it has been called, is infinitely greater. Humus is said to be able to absorb from 40 to 60 per cent. of water, and to hold it very tenaciously. This is from two to three times as much as the most porous dead mineral soil is capable of absorbing. We all know that in times of heavy rain it is infinitely rarely that we see water lying in pools on the surface of cultivated soil, whereas it soon collects on roadways and paths, which are made of dead mineral matter. The tenacity with which mould retains water is due to the fact that the water is absorbed into the interior of millions of vegetable cells, and is not merely held by capillary attraction in the interstices between small mineral particles. It is the swelling of individual cells which forms so effectual a barrier to the passage of bacteria.

Not only water, but everything else when thrown upon the soil, disappears sooner or later. Such things as pieces of wood, or leather, about the toughest of organic materials, become softened and permeated by fungoid growth, and finally crumble away. In some parts of the country, rags of all kinds are largely used for manure. Through the autumn and winter these may be seen lying on the surface, but when in spring the tilling of the land goes forward, and the fermentation of the soil commences, the coarsest of these rags disappear. If wood, leather, and rags disappear, leaves and animal excrement disappear, as we all know, far more readily. The disintegration is forwarded by birds, insects, worms, and their allies, and what was the excrement of a large animal becomes, as it were, the excrement of many small ones, until finally, by the action of saprophytic fungi, these organic matters become fertile 'humus,' which is the only *permanent* source of wealth in any country, the source whence we derive all the materials for our food and clothing.

The question whether among the bacteria which are found in the soil some may not be hurtful to mankind, is a question of great interest and importance. If disease-causing organisms find their way into the soil, may they not multiply or at least continue to live, and then prove a danger to health? There can be no doubt that pathogenic organisms do exist in the soil, but their power for harm would seem to be practically very small indeed; and to regard the soil as dangerous because some pathogenic organisms may lurk in it, would be about as rational as it would be to condemn vegetable food because of the occasional dangers of hemlock, aconite, or the deadly nightshade. It is well known that if soil be inoculated into some of the lower animals, such as guinea-pigs, fatal results will follow from malignant œdema and tetanus; and also that earth, and especially street-mud, if ground into wounds in the human subject, may cause similar diseases, and the death of the victim. It is equally well known, however, that the workers of the soil, agricultural labourers and gardeners, are amongst the healthiest classes of the community, and that they are not credited with any diseases which are special to their calling. The disappearance of malaria (a real soil-poison) when land is drained and tilled is a fact which is interesting in this connection. It seems to be a fact that the great doctrine of '*the survival of the fittest*' holds good for microbes in the soil, as for all other organised things everywhere; and that organisms which flourish in the human body, languish and cease to multiply in the soil, where the conditions are unsuited for their multiplication or even for their survival. They get overgrown by saprophytic microbes, and even if they do not die the risk of their finding their way into the ground water is practically nil, for we have seen that humus is the best of filters.

The life-history of at least one microbe, which undoubtedly flourishes in the human intestine, has been very carefully studied by many observers, and it may profitably occupy our attention for a time. This is the so-called spirillum of Asiatic cholera, the 'comma bacillus' of Koch, of which we have heard so much, and which is now generally accepted as being the *Causa causans* of the disease. The subject is brought forward merely as the life history of a microbe which undoubtedly flourishes in the human intestine, and has not been found except in association with a deadly disease. This microbe, which has been met with exclusively in the dejecta of cholera patients, is easily cultivated on gelatine or potatoes, in neutralised meat infusion, on blood serum, and in milk, its growth being unaccompanied by any disagreeable odour. Growth ceases when the infusions become very dilute, and in water growth only takes place at the margin where there is an accumulation of nutrient material. Growth is able to take place with a very limited supply of oxygen, as is shown by its multiplication in the intestine, and it is most active when the temperature is high—30° to 40° C. Koch has made the very interesting observation that comma bacilli die very rapidly when dried. A cultivation if spread out upon glass and exposed to the ordinary temperature is dead and incapable of further multiplication in a very few hours. Hence it is inferred that no living comma bacillus can exist in dust, and that the transport of *living* comma bacilli through the air is impossible.

Another factor very unfavourable to the growth of comma bacilli is the presence of saprophytes in large quantities; under these circumstances they are overpowered, and die out. 'If the saprophytes are in excess in the first instance, or if the sum total of the conditions of life are not very favourable to the comma bacilli, the latter do not multiply at all, but the saprophytic bacteria lead rapidly to the death of the comma bacilli present, either by using up the nutrient material or by producing poisonous products' (Flügge). If, however, the bacilli be kept moist in the absence of saprophytes, they may be kept alive for months. Low temperature (freezing) does not kill them, but merely suspends their vitality; temperature over 60° C. soon kills them.

If the bacilli find their way into pure running water, or wells of 'pure' water, it is probable that multiplication never occurs. In the case of stagnant water, however, in the bilge-water of ships, in the water in harbours, which is often extremely dirty, it is probable that the comma bacilli may retain their vitality for a much longer time; and in the case of a tank in India, 'where the small amount of water was not only employed for bathing, drinking and cooking, but also for washing the linen and for the reception of the contents of the water-closets, Koch was able to demonstrate such a large number of comma bacilli that it seemed likely they had multiplied to a great extent in the tank, and that their presence was in all probability the source of infection of a number of cases of cholera which occurred at a later period among those persons who lived in the neighbourhood' (Flügge).

Supposing comma bacilli to exist in dejecta, what is the best way to stop their multiplication and accidental passage into drinking water? Clearly to dry them and place them with other saprophytes. If they be buried in the upper layer of vegetable mould the sun will dry them; or even if it be raining the living filter will stop their passage downwards. The growth of saprophytes will kill them; and if the ground be cultivated, the comma bacilli will be destroyed and nitrified, and pass upwards into the crop, and not downwards into the wells. If, on the other hand, the dejecta be mixed with water, and be taken in an impermeable pipe through the living humus of the surface, to the dead mineral subsoil where the sun does not reach to dry them, and where saprophytes to eat them up exist not, the danger of their finding their way through interstices and crevices into drinking water appears to be very great indeed, especially if the dirty water be in a cesspool which leaks *under pressure*, as is explained elsewhere.

That the under layers of the soil are a very inefficient barrier against filth contamination has been demonstrated in all our large towns, and especially in London. In that city the lower rooms of the houses are almost universally below the level of the street, and the house drains leave the house at the lowest point to reach the sewer at a lower level still. As underground drains, however well laid, are sure to leak in time, their contents escape; and water continually escaping at one point is sure to work a channel for itself, and take its natural course to the nearest stream or well. Still more is this sure to happen if the house drain leads to a cesspool, a contrivance which necessity invented as soon as we had water under pressure, and began to use it as our only scavenger.

In London, a city renowned for its innumerable wells, we have had to close every one of them, and as the excessive dirtiness of the air makes rain-water not available for domestic purposes, we have become absolutely dependent upon the water companies, and it is only quite recently that the public has become alive to the fact that the causes which poisoned the surface wells are equally poisoning the Thames and the Lea, and the other sources of London water. No thinking being can feel easy about the London water-supply, and it is to be hoped that some day the public mind will be roused to an appreciation of the fact that if we want pure water we must make some serious attempt not to foul our wells and streams.

One cannot but feel that in our sanitary arrangements we have not sufficiently distinguished between the living mould of the surface and the dead earth of the subsoil. The living mould is our only efficient scavenger, which thrives and grows fat upon every kind of organic refuse; our only efficient filter, a filter which swells and offers an impassable barrier to infective particles, a filter which affords a sure protection to our surface wells. When we perforate the living humus with a pipe, and take our dirty water to the subsoil, we, as it were, prick a hole in our filter, and every chemist knows what that means.

In order to keep the soil healthy, to keep up its appetite for dirt and its power of digestion, the only thing necessary is tillage. Well-cultivated soil, which is compelled to produce good crops, has never yet been convicted of causing any danger to health.

Sanitation is purely an agricultural question, and in the country, where every cottage has, or should have, its patch of garden, there ought to be no difficulty in the daily removal of refuse from the house, and in applying it to agricultural purposes, without any risk of contaminating the water-supply. Given the patch of garden, the only thing necessary to bring about this, the only complete form of sanitation, is the will to do it—the will, that is, to do a profit to one's self, without the possibility of damaging one's neighbour. This, unfortunately, is rarely forthcoming, in spite of the Christian religion and the Education Act, and we go on, even in country places, polluting our streams and wells, with our minds agitated, as well they may be, as to when our water will become too poisonous to drink, and where we shall turn for a pure supply in the future.

Sanitation is a purely agricultural and biological question. It is not an engineering question, and it is not a chemical question, and the more of engineering and chemistry we apply to sanitation the more difficult is the purifying agriculture. This, at least, has been the practical result in this country.

The only engineering implements which the cottager with a bit of garden requires for his sanitation are a water-pot and a spade, and if his garden be an allotment away from the cottage, a wheelbarrow may become necessary. The cottager, to whom the produce of his bit of land is a matter of consequence, will endeavour to fertilise as much land as possible with the organic refuse at his disposal, and as long as this endeavour is made there need be no fear of failure, either from the agricultural or sanitary point of view. When, however, an engineer, by means of water under pressure, has collected the organic refuse of a province at one spot, has diluted it a thousandfold, and endeavours to submit it to a mock purification by means of the least amount of land possible, failure is inevitable, both in the agricultural and sanitary sense. It was in 1848 that the advice to 'drain' was tendered with a light heart by the pioneers of modern sanitation, who thought it would be an easy thing to purify the sewage and make a profit from it. The Thames, the Liffey, the Clyde, the Mersey, and the Irwell are a standing testimony to the failure of these great engineering schemes, and one of the last engineering schemes put forward with regard to the sewage of London, viz. to convey it all to the Essex coast and cast it into the sea, is not only a most lame and impotent conclusion, quite unanticipated by the pioneers of '48, but it is an experiment which, like our previous experiments, may be productive of unforeseen results.

The engineer of the present day, when dealing with sewage, appears to think that one may ' as well be hung for a sheep as a lamb,' and he is ever ready to tender the advice that 'if you are going to make a mess, it is well to make a big one.' It is quite characteristic that this last scheme for dealing with the London sewage contemplated dealing not only with the material which is collected by our present system of sewers, but proposed to take that of other and adjacent systems as well.

The people of Berlin have in this respect shown themselves wiser than the people of London, because they have taken their sewage to several points instead of collecting it all at one spot.

The panacea for all sanitary ills has been and still is 'drainage,' and the only scavenger that is in favour is water, notwithstanding the fact that sanitation by water has for its main characteristic 'incompleteness.' The work is begun and never finished. Our houses are flushed, but we pay for it by fouling every natural source of pure water, whether river or surface well. If there come an outbreak of typhoid, we, as often as not, find the 'drains' are to blame, but, as a matter of fact, we prescribe 'more drains' as the remedy.

The author has asked his friend and former pupil, Mr. F. W. Wells, M.B., to go through the official reports which have emanated from Whitehall since 1856, and make an abstract of the chief outbreaks of typhoid fever in this country which have been reported by the medical officers of the Privy Council and the Local Government Board. This Mr. Wells has done in a most painstaking and methodical manner, but the tables which he has constructed have been omitted as unsuited to the present volume. A perusal of these tables showed that there is one factor common to all these outbreaks, viz. the mixing of excremental matters with water. This mixture generally leaks to the well or rivulet, or water-pipe which supplies the drinking water, which water has not unfrequently been sold under the name of milk, and the result is an outbreak of typhoid. Or the mixture putrefies in a cesspool or sewer, and the gases, finding an entrance to our houses, cause an outbreak of typhoid. There is no doubt whatever that whenever excrement is mixed with water we are in danger of typhoid. Typhoid was not recognised in this country until the water-closet became common. We, doubtless, manufactured typhoid in a retail fashion in old days, but with the invention of the water-closet we unconsciously embarked in a wholesale business.

We had not been at this work many years before we recognised that the water-closet poisoned all sources of water. We have had to go far afield for drinking water, and the result has been that, as we have left off consuming the springs which we have wilfully poisoned, the amount of typhoid has somewhat abated. When the more remote sources of water get poisoned in their turn—as with our increasing population, and our methods of sanitation, is inevitable—the present comparative abatement must, one would fear, cease.

The foregoing observations apply, be it observed, to cholera equally with typhoid.

It is comparatively recently that we have learnt to recognise the dangers which result from the putrefaction of a mixture of excrement and water in a sewer or cesspool. The ingenuity of sanitary engineers has been exercised to save us from these dangers, and they have given us what they are pleased to call self-cleansing sewers, innumerable forms of trap, endless methods of ventilation, and disconnection on scientific lines, until the medical officer of health is expected to have at his fingers' ends all the knowledge of a patent agent and a plumber's foreman. If apparatus never wore out, if ventilators never got stopped up, if traps never got unsealed by leakage, evaporation, or other cause, one might feel secure against the enemy which is ever at our gates, provided the study of Bacteriology did not lead us to recognise that a few feet of filthy pipe may be as dangerous as a mile, and that a trap may possibly serve, especially in hot weather and when the family is away, as a most efficient 'cultivating chamber.'

It is commonly urged by those who defend our present methods of sanitation that, as we must of necessity provide some channel for the escape of slops from our houses, it is false economy not to make these channels carry everything; or, in other words, that, as sewers are a necessity, there is no harm in making them a bigger nuisance than they necessarily must be. It is difficult for the author to follow this argument, and he would submit some reasons why every effort should be made to keep excremental matters out of the sewers.

1. Excrement is the only ingredient of sewage against which dangerous infective properties have been proved again and again. It is the ingredient which, when mixed with water, finds its way to our drinking water and causes typhoid and cholera. Sewage without excremental matters is, doubtless, offensive, and is probably unwholesome in many ways, but it stands in the position of a 'suspect' rather than that of an habitual criminal against whom no end of previous convictions have been proved.

2. If excremental matters were stopped out of our house drains we could, in country places, often have recourse to the old practice of allowing our household slops to run in open gutters, concerning the ventilation of which there could be no doubt, and the gutters might be subjected to the wholesome discipline of a broom and the purifying influences of sunlight and drying winds.

3. If excremental matters be stopped out of the house drains, the total volume of sewage to be dealt with would be diminished by at least one-fifth, and this surely is a great gain. We should deprive the sewage of just those ingredients which are most troublesome to the sewage farmer by clogging the pores of the ground, and we should leave the sewage very 'thin' and admirably suited for downward filtration. It seems to be an acknowledged fact that, for the application of sewage to the land, the more watery it is and the more completely solid matters are strained out of it, the simpler and more satisfactory the processes become.

4. Another class of objections which has been made to the exclusion of solid excrement from house drains has reference to the so-called 'manurial value' of sewage and its constituents—that excreta without the total urine are of low manurial value, and that the stopping of excreta out of the sewers lowers the manurial value of the sewage. 'Manurial value' is a term used by chemists to express the amount of nitrogen that may be present. Now one does not doubt the ability of chemists to make a quantitative estimation of nitrogen, nor their power of informing farmers of the extent to which they may or may not have been cheated when they purchase artificial manures, but it is evident that the real practical manurial value depends, not only upon the amount of

plant-food present, but also upon whether the plant-food is present in a form in which it can be digested and exhaustively utilised by the plant. For the latter information, which is of the highest importance, the author would sooner apply to a practical farmer or gardener than a chemist.

A chemist, for instance, who had regard to his analyses and nothing else, might tell us that nut-shells had a certain dietetic value; but ordinary men and monkeys know better than that.

He might tell us that gin was richer in certain dietetic ingredients than ginger-beer, but we know that ginger-beer is the better article of diet.

Again, guano has a far higher manurial value than 'rich garden mould'—such as is got by mixing earth with organic refuse; but if we do not dilute our guano to the same level, so to say, as our rich garden mould we may kill our plants. To declare that rich garden mould is of low manurial value is absurd, because we know that in it plants of all kinds reach the highest development which is attainable. Farmers and market gardeners will tell you that artificial manures have 'got no bottom in them,' that their use is, so to say, a speculation; and if climatic conditions are unfavourable when the artificials are applied, the money spent on them is lost for ever. With organic refuse, however, the case is entirely different, and the effect of the application of organic matter, especially of human origin, to the soil is plainly discernible for three or four years. Solid organic matter is little liable to be washed away, it nitrifies slowly, and doles out the nitrates to the roots of the plants in proportion as they are needed.

It is necessary to say, emphatically, that the manurial value of human excrement is enormous, and that it produces all kinds of fruits, flowers, and vegetables in the highest perfection. Speaking from a practical experience of ten years, my belief is that soil cannot be made more fertile than by mixing it with solid excremental matter.

It is quite true, no doubt, that the manurial value of urine is very great, but being fluid it is not so easily retained at the spot where the agriculturist wants it; and we know that when fresh and undiluted it is very dangerous to herbage. The fact is that plants absorb their nutriment from very dilute solutions; and it has been found that a fluid containing about ·2 per cent. of solids is the *optimum* for plant culture. Ordinary urine, therefore, which contains 4 per cent. is twenty times too strong; but if it be applied to the soil in its state of optimum dilution, much of the liquid will necessarily soak out of the reach of the roots.

Manurial value is a practical matter rather than a chemical problem, and there is no doubt whatever that those who assert the manurial value of earth-closet manure to be low are making a very serious practical mistake; and there is no doubt that arguments based on the theoretical manurial value of sewage as a whole, or of its several ingredients, are worthless in helping us to decide whether it is advisable or otherwise to keep solid matters out of the drains.

What use is there in discussing the 'manurial value' of sewage in the face of the deliberate declaration of that eminent agriculturist, Mr. Clare Sewell Read, made a few months since in the 'Journal of the Royal Agricultural Society'? 'Sewage,' says Mr. Read, ' has come to be regarded by all sensible people simply as a nuisance to be got rid of.' And he goes on to state that, owing to the unmanageable quantities of water which have to be dealt with, sewage is ruinous to all grain crops and all other farm crops except rye-grass.

The composition of sewage as it flows from towns is so doubtful, and must be so variable, that no sensible man would let it run over his farm. Chemicals and antiseptics are very abundant at the present day, and they are very largely used to lessen the dangers which are inherent in our present system of sanitation. Antiseptics, however, which stop the growth of putrefactive microbes, also check the growth of nitrifying organisms, and are deadly poison to plants. All town sewage is liable to contain dangerous chemicals which must render the 'manurial value' a very minus quantity, the presence of nitrogen notwithstanding.

As it is idle to discuss the theoretical manurial value of a practical nuisance which no sane farmer would take as a gift, it is imperative for us to discover means, if possible, by which those ingredients of sewage which have great enriching power for the soil may be saved for the benefit of the cultivator and consumer.

From every point of view—scientific, sanitary, moral, economic—the author feels strongly that dwellers in the country should take warning by the towns. They should revert to the cleanly and decent habits of our forefathers, and keep the sanitary offices away from the main structure of the house, and not, as is the filthy custom of the present day, bring them almost into the bedrooms. They should keep solid matters out of the house drains, and see that they are decently buried in the living earth every day, and they should replace the drains by gutters and filter all the household slops by applying them to the *top* of a different piece of cultivated ground every day. Whether an ordinary watering-pot, or a tank upon wheels drawn by a horse, be necessary for accomplishing this latter object will depend upon the size of the establishment ; but only those who have systematically pursued this plan can know the vigour which is imparted to hedge-rows, shrubberies, fruit trees, or forest trees by a tolerably frequent dose of household slops. There is no difficulty in doing this, provided the will be present—the will that is to combine your duty towards your neighbour with an act which is profitable to yourself.

Finally, to dwellers in the country, whether squires who are the owners of broad acres, or occupants of modest villas with a garden, or still more, to cottagers with an allotment, where it ought to be, round the cottage, the following principles of action are recommended :—

1. That sewage, being a nuisance, although a necessity, it is to our interest not unnecessarily to increase its quantity or its offensiveness.

2. That excrement should be kept out of the drains, for by doing this the putrefaction of the solid is prevented and the purification of the liquid by filtration through the earth is effected with ease which is proportionate to the thinness of the fluid.

3. That all solid matter should be removed every day from the immediate neighbourhood of the house and buried in the top layer of cultivated ground. Household slops should be poured on to the surface of the garden, and the mistake of attempting what is known as 'subsoil irrigation' must not be made.

If these directions be followed, it is evident that by no possibility can one be troubled by sewer gas, and it would probably be no longer dangerous to drink from surface wells.

The practical experience of the author in dealing

with sewage in the manner which is here advocated will be dealt with at some length in a subsequent chapter.

Although the author is addressing himself to dwellers in the country, he would like to say to town dwellers that complete sanitation is impossible, unless cultivated land be brought into tolerably close relationship with the dwelling. At present our sanitary arrangements are magnificently begun, and seldom completed, and while we almost uniformly leave a most dangerous loose end to our sanitary measures, we shut our eyes to it, and blow the trumpet of self-satisfaction as if the sanitary millennium had begun. The Allotment Act, as affording an outlet for organic refuse, ought not to be without its effect upon sanitation, and it is to be hoped that the masses will some day wake up to the great importance, from the moral and sanitary standpoint, of providing every dwelling with an adequate outlet. As things go at present, there is very little doubt that the agricultural labourer with his cottage and garden and 12s. a week is infinitely better off than the town artisan on 25s., who pays dearly for pigging it in overcrowded rooms, in which a cleanly and decent existence is impossible.

In a recent volume of the Transactions of the Sanitary Institute is a very interesting paper by Dr. Sykes, who quotes Dr. Corfield, who, in his turn, is quoting Sir Henry Acland, to the effect that the disappearance of the great cities of antiquity was due to pestilence rather than war. We must all admit the possibility of such an assumption, and certainly no one can ponder upon the disappearance of Egyptian, Babylonian, Assyrian, Greek, and Roman civilisation without speculating upon the cause, and without applying the lesson to ourselves, and asking, how much longer is our British civilisation to continue?

Nationalities seem as mortal as the individuals which compose them.

If great nations are destroyed by neglect of sanitary laws, and if prolonged national life is indicative of sound sanitary measures, there is at least one race upon the globe which is worthy of profound study by all who concern themselves with public health. This race is the Chinese, who have seen all the nations of antiquity in and out, who were probably a great people in the days of Moses and before, and whose thrifty myriads are even now successfully contending with the Anglo-Saxon race in America and Australasia. The Chinese, as is well known, have had to contend with national calamities of a most stupendous kind. In our own days we hear of floods and famines which claim their millions of victims, and yet the race continues to increase in such a way, and to overflow its natural boundaries to such an extent, that it is certain, even without the exact returns of a Registrar-General, that the birth-rate must very considerably exceed the death-rate, and must have done so in an average way during the three or four thousand years that the Chinese nation has existed.

There is little doubt that, unless we mend our ways, the Chinese will see us out, as they have seen the other great nations of the world out, and the reason for this belief is obvious. The Chinese are the most thrifty nation in the world. In China nothing is wasted, and all organic refuse is ultimately returned to the soil. Even the dead bodies of the Chinese who die abroad are returned to China for burial. Agriculture is, in China, a sacred duty, and the Chinese have got a firm grasp of the elementary principle that, if the fertility of the earth is to be maintained, we must constantly replenish it. The nineteenth volume of the Health Exhibition literature contains a most interesting series of papers on China, by Surgeon-General Gordon, Mr. Hippisley, and Dr. Dudgeon, of Pekin. The papers by Dr. Dudgeon are especially worthy of study, for many years of residence among the Chinese have impressed him with the fact that we have much to learn from them.

The question of our duty to the soil is fundamental in sanitary matters. If we starve the soil and turn our fertilising materials into the sea, we may rid ourselves (though this is doubtful) of filth diseases for a time; but it is by no means doubtful that we shall ultimately replace filth diseases by those diseases that are bred of starvation. How soon this will happen no one can say, but that it will happen eventually seems to me as certain as is the axiom 'Ex nihilo nihil fit.' Do not let us commit the great blunder, when dealing with this national question, of forgetting that the life of a nation ought to be measured by centuries; do not let us make a suicidal use of a paltry fifty years' statistics, and because the figures of the last decennium happen to be favourable, conclude therefrom that all our sanitary principles are right.

Perhaps someone will say, ' How ridiculous to hold up the Chinese as an example ! The Chinese masses are acknowledged to be exceptionally filthy in their customs and habits.' This, perhaps, is true, but I am sure that the reader will not make the error of confounding principles with details. The Chinese principle of returning all organic refuse to the soil is, there can be no doubt, absolutely sound. The Chinese details may be filthy and susceptible of improvement. In this country the details of our domestic sanitation are refined, elegant, and ingenious. It is the principle subserved by these details which is absolutely rotten. The main problem of sanitation is to cleanse the dwelling *day by* day, without fostering starvation. This can only be done by returning all organic refuse to the soil, and the perfecting of the details by which this duty is to be done is the most important work of the modern sanitarian.

This question is a national one, and concerns us all. Every country squire ought, in these matters, to set a good example to his tenants. If he does not set the example of increasing the fertility of the soil by the daily addition to it of all the organic refuse of his country mansion, he cannot command our sympathy when he goes without his full rent. If a landowner embarks on a great building scheme, he ought to keep the sanitation in his own hands. If a well-known landowner had done this—if he had preserved his autonomy on his own estate, and if he had, by a rational use of the railway, transferred the daily scavengings of his valuable City estate to his broad acres in Bedfordshire, perhaps his right-of-way on his London estate would not have been contested, and perhaps he would not have been obliged to remit 25 per cent. of his Bedfordshire rental. As it is, he allowed the vestry to do his sanitation for him, and by so doing lost his autonomy. Who can see how far the process of confiscation which has set in will ultimately reach?

This question has an immediate personal interest for all who derive their income from the soil. The clergy would do well to enforce by example as well as by precept the old injunction, to 'replenish the earth and subdue it.' If they do not, they must expect to go without their tithes. Improvement in this direction is only to be attained by rousing the public conscience. So soon as the majority of individuals is impressed with the fact that it is wicked to foul our streams and starve the soil, and that our individual responsibility does not end, even though the fouling and starving be done by a 'Board,' so much the better will it be for the public health and national wealth. Parliament has compelled us to hand over our responsibilities to public authorities, with the consequence that the individual has lost his liberty and independence, and is drifting into a condition of sanitary imbecility. Let us not forget that the present state of our rivers is the direct result of Acts of Parliament. Let us not forget that Parliament, which wasted its time and our money in passing that most inoperative of all Acts, the 'Rivers Pollution Act,' scavenges its own palace direct into the Thames; as though Imperial Parliament could hand over its responsibilities to a Local Board! It is hardly credible that such a condition of things could exist outside the libretto of a comic opera.

A respect for the purity of water should be enforced in our Board schools and churches; and that powerful party in the State—the Temperance party—would do well to devote some of its energies towards ensuring that the beverage which it champions should be in all places a safe one to drink. As it is, one has only to walk about the country to see that our streams and rivulets are universally regarded as receptacles for rubbish and impurities of every kind.

This question is a national question of the first importance. A nation that fouls its streams and starves its soil is in danger of poisoning and inanition. A nation which imports a great part of its food and a great part of its manure, and systematically and by Act of Parliament throws all its organic refuse into the sea, is undoubtedly living on its capital. Our capital just now is undoubtedly considerable, but we are in a fair way to run through it; and when we have done so who can forecast the future?

In connection with the power of 'The Living Earth' to deal with organic fluids, a very interesting series of experiments has been made at the author's suggestion by Mr. Wells on the filtration of urine through earth, and what follows is an abstract of Mr. Wells's report on the matter.

A conical metal vessel (fig. 1), capable of containing about 40 lb. of dry earth, and perforated at the apex, was used as a filter, and to this was added day by day an amount of urine averaging about half a pint. In the first three experiments, made between July 1890 and February 1891, fifty-nine separate analyses were made of the urine committed to the filters, involving the estimation of urea, sodium chloride, organic and mineral residues, and in subsequent experiments the averages of these analyses have been regarded as sufficiently accurate for purposes of comparison between urine and filtrate. The experiments are arranged in three divisions:—
(1) Experiments made with mould from Gower Street;
(2) experiments made with mould from Brondesbury;
(3) experiments made with mould that had been previously used.

In the first experiment the following results were obtained: 200 oz. of urine were added to the filter, of which 73 oz. appeared as filtrate, the remaining 127 oz. representing what was lost by evaporation *plus* what remained in the filter. The average specific gravity of the urine was 1021·4, that of the filtrate 1011. The total solids of the urine averaged 4·44 grammes per cent., of which 3·45 per cent. were organic and 0·99 per cent. inorganic, while the total solids of the filtrate averaged 1·78 gramme per cent., of which 1·069 was organic and 0·710 inorganic. The chlorides in the urine averaged 1·09 gramme per cent., and in the filtrate 0·67 gramme per cent. The total solids of the urine added would be nearly 9 oz., and the total solids of filtrate obtained about $1\frac{1}{3}$ oz., so that $7\frac{2}{3}$ oz. of solid residue were left in the filter. The

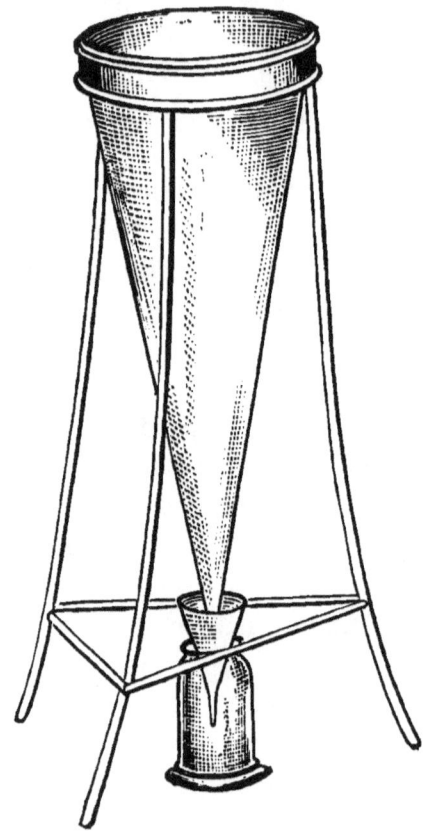

Fig. 1.

appropriation of these $7\frac{2}{3}$ oz. of residue by the filter will be again referred to.

The surface of the filter, when the nose was closely applied, smelt distinctly ammoniacal; but the filtrate, although of deep colour, could be evaporated to dryness without offensive smell. The filtrate showed no tendency to putrefy, nor did a few drops of it, added to sterilised urine in a tube, set up putrefaction. Next as to the experiments made with a different humus, two of which were conducted in the garden at Brondesbury. In all 570 oz. of urine were added to the filters in equal additions of 10 oz. daily. Of this 228 oz. were recovered as filtrate. The average specific gravity of the urine was 1020·3, and of the filtrate, 1006. The urine contained an average of 2·34 grammes per cent. of urea; the filtrate, 0·15 gramme per cent., an amount which, allowing for possible error in the hypobromite method of estimation, may be disregarded. The average chlorides of the urine amounted to 1·35 gramme per cent., and of the filtrate to 0·33 gramme per cent. Thus of about 14 oz. (13·9) of urea added to the filter, $12\frac{1}{3}$ (12·3) oz. did not appear in the filtrate, and of 8 oz. (8·03) of chloride added $6\frac{1}{2}$ oz. (6·6) were unaccounted for. Nor do these results represent the best that have been obtained. Thus, in the last experiment made at Brondesbury, the first pint and a half of filtrate was almost colourless, of specific gravity 1003; chlorides, 0·2 gramme per cent.; and urea, 0·13 gramme per cent.—an amount which, as already pointed out, may be disregarded.

None of the filtrates, after standing for weeks exposed to the air, underwent any putrefactive change, nor did any of them give off offensive smell when evaporated to dryness. In colour they have varied from the almost colourless filtrate above to a yellow or deep reddish-brown. In order to ascertain whether the colouring matter was of a urinary nature, part of the filtrate obtained from the first experiment was sent to Dr. MacMunn, of Wolverhampton, who kindly examined it, and has furnished the following report of his examination:—

'*Colour.*—A deep yellow, with orange tinge by gaslight, brownish tint by daylight.

'*Spectroscopic examination.*—A layer 40 mm. deep absorbed a little of violet end of spectrum. No detached band in any depth. With HNO_3: No band; colour diminished. With H_2SO_4 in abundance: Colour almost discharged. By Jaffé's test (HCl and bleaching salt): Colour discharged. With NaHO: A precipitate carrying down all colouring matter: filtrate clear, showing no bands; some ammonia set free. With $NH_4OH + ZnCl_2$: Brownish precipitate; filtrate clear, and gives no fluorescence.

'*Result.*—A total absence of normal urinary pigment and chromogen—that is, no urobilin, no indican, &c. The colour is probably due to the presence of some humus substance derived from the earth through which it has been filtered.'

The third group of experiments were made with mould that had been previously used, an interval of five months having elapsed, during which the mould had remained intact. One hundred and nine ounces of urine were added, of average specific gravity 1024·3, with the following percentage analysis:—Urea, 2·6 grammes; chlorides, 1·2 gramme; residue, 5·48 grammes, being organic 4·12, mineral 1·35. Twenty-four ounces of filtrate were recaptured, of a uniform specific gravity 1035, with urea 0·2 gramme per cent. Thus the power of the filter to change the urea was as great as ever. Of chlorides there were 1·5 per cent., and residue 6·07 per cent., being organic 3·37 and mineral 2·70 per cent.—that is, the mineral residue of the filtrate was exactly double the amount of that in the urine; and it was evidently not due to chlorides, the respective average percentages being—urine, 1·2 gramme; filtrate, 1·5 gramme. It seems probable that the 7⅓ oz. of residue left in the filter during the five months' interval has to a great extent nitrified and become soluble. Given an ammoniacal solution and garden soil, we have the readiest means of inducing nitrification. And, testing the filtrate qualitatively with ferrous sulphate and sul-

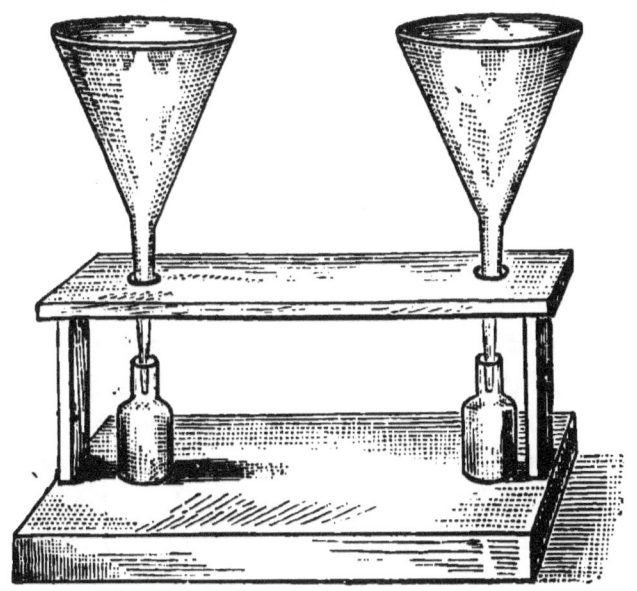

Fig. 2.

phuric acid, there was abundant evidence of the presence of nitrates.

Allusion may also be made to an experiment intended to ascertain what difference in results might be obtained from a water and a urine filtration through mould. Two glass funnels (fig. 2) were employed, each of about four pints capacity, and were filled with mould. To one was added during a period of twenty days, in regular amounts, 1½ oz. of tap water, and to the other an equal amount of urine. Rye-grass seed was, moreover, sown on each filter surface. The water-filter produced 7½ oz. filtrate, and developed an excellent crop of rye-grass; the urine filter produced 5½ oz. of filtrate and no grass. The daily addition of *water* was followed, after the filtrate had once commenced to appear, by an almost *immediate* dropping of filtrate; *the urine tarried much longer in the mould*. That no grass would grow on the mould to which urine was being daily added was to be expected, ordinary urine being about twenty times too strong for plant culture. The filtrate obtained from the urine filter was odourless, clear, of pale yellow colour, specific gravity 1006, containing about 0·5 gramme per cent. of urea (? or other nitrogenous bodies), and yielding no offensive smell on evaporation to dryness. The water filtrate was of faint yellow colour, and of specific gravity 1000. Finally, a few experiments were made as to the fertility of the mould from the filters. In May 1891 a mixture of mould that had been used for filtration experiments (one-half of which had, moreover, been sterilised by heat) was sent to Andover, and there twenty cuttings of common yellow calceolaria were placed in pots as follows: Ten in garden soil from the garden at Andover; ten in the London mould. The pots were placed in the open. Here is the result: In the ten pots of Andover soil the calceolarias had done fairly well, and each pot had grown an abundant crop of weeds. In the ten pots of London earth the calceolarias were all dead, and there was not a single weed visible. They were the embodiment of absolute barrenness. Corresponding portions of mould were used for urine filtration, and yielded a filtrate of specific gravity 1041, with a large percentage of urea. In no case has grass grown on the surface of a filter in use; but, on mould that has remained for several months exposed to the light and air in the garden after usage in the filters, grass has been cultivated most readily, yielding a better crop than that produced by unused mould.

Although these experiments are manifestly incomplete in some particulars, they show very conclusively how great is the power of the earth to deal not only with organic solids but also with organic liquids, such as urine. The author is not aware that any experiments similar to the above have been published, and attention may be directed to four facts: 1. That the filtrate obtained from the filtration of urine through fresh earth was always a much thinner fluid than the urine added, the bulk of the solids dissolved in the urine having been left behind in the earth. 2. The filtrate showed no tendency to putrefy, and certainly contained no putrefactive organisms, for it was shown to be incapable of starting putrefactive changes in urine which had been previously sterilised. 3. The filtrate could in all cases be evaporated to dryness without giving off offensive odours, offering in this particular a great contrast to pure urine, which invariably emits most disgusting odours when evaporated to dryness. 4. The organic residue left in the earth apparently underwent nitrification in course of time, but it was noteworthy that earth which had been used for the filtration of relatively a very large amount of urine was in all cases barren until it had been exposed to air and rain. After such exposure, however, its fertility appeared to be of a high order. The practical importance of this matter to sanitarians and agriculturists is very great. In London, with its 4,000,000 inhabitants and endless throng of visitors, it is probable that 10,000,000 pints of urine are daily run to waste, and it must be remembered that these 10,000,000 pints contain about 200 tons weight of most valuable manure.

It is interesting to note that humus from different localities seems to differ in its power of dealing with urine. The humus from Gower Street (which is largely adulterated with London 'blacks') seems to be undoubtedly inferior to that of Brondesbury. The best of the Brondesbury results, in which the filtrate had a specific gravity of only 1003, and contained no appreciable amount of organic matter, was certainly very remarkable.

CHAPTER V

THE HOUSE

BY those who are content to live simply, the conditions of healthy living are easily attained.

A cottage (*ornée*, if you like) built on a slight eminence or on the slope of a hill, well exposed to the sun, and with a quarter of an acre of land, is no Utopian idea. The idea is capable, be it observed, of contraction or expansion, but if the individual is to be *quite* independent of others for his healthy surrounding, a small plot of garden ground is absolutely essential.

The foundations of this modest house must be solid and dry, the walls thick, the windows big, and the roof watertight.

There must be neither cesspools nor sewer, which are great and acknowledged causes of sickness. No filth or refuse must be allowed to accumulate, but it must be returned *every day* to the soil.

The rainwater which falls upon the roof should be kept for washing and cooking, if not for drinking; and it may be remarked parenthetically that there is no reason why rainwater should not be collected in an ornamental vessel rather than the dirty-looking water-butt or hideous tank which is now too much in vogue. It is far better to have an ornamental water vessel in a prominent place, where it can be easily inspected, than a hideous receptacle poked away in some odd corner where it is too often forgotten.

There are one or two points in the construction of dwellings which are of importance, and which do not receive sufficient attention from architects.

One is to have a maximum amount of light precisely in those places where filth is most likely to accumulate, in order that the filth and dirt may be easily detected.

Another is to let waste water be discharged at the highest level possible, in order that its filtration by gravitation through the earth, or any other suitable filter, may be easily arranged.

A third is not to allow the staircase to become a channel whereby the used air of the ground floor may float up to the first floor. Staircases should be shut off from the ground-floor apartments by means of a door.

It will be noted that what has been sketched is within the reach of very moderate means. It is probably too simple, and will not be approved of by those who have longings after more pretentious residences and more artificial modes of existence.

Health depends upon our obeying the beneficent laws of nature; and the rule of nature which most affects our health is this, *that all refuse matter shall be restored without delay to our mother earth*, who will receive it gratefully, and give back a dividend. The greater number of our sanitary troubles are due to the neglect of this law. It will be noted that it is very cheap to live healthily. In the dwelling I have supposed there would be no sewer-rate, no water-rate, no plumbers' bills, very little sickness, and a good return from the well-cultivated and well-nourished garden.

A distinguished physician and sanitarian has drawn a picture of a city in which the laws of health are strictly obeyed; he called his imaginary city the City of Hygeia. A City of Hygeia is hard to attain, and would involve immense expenditure and endless watching. A healthy cottage (call it Hygeia if you like) is easily attained, and is the cheapest dwelling imaginable for a civilised man.

Without cultivation of the soil there can be no high standard of health. The gardener and the farmer are, so to say, the right-hand men of the sanitarian. What the householder wants to be rid of, the tiller of the soil is ready to take. If the householder tills a bit of soil for himself all difficulties are at an end; if he do not, his difficulties begin. If the refuse of a single house be put on its own garden, there is no difficulty. If the refuse of a group of houses be taken to a neighbouring field or garden, the difficulty is slight, but the return is less, because the expense of transport has to be borne, and it must be remembered that refuse matter will not bear the expense of much carriage. If the refuse of a big town is taken to one spot, the delay in collection and other difficulties increase, and it is found that the expense of transport eats up the profit. It is evident that refuse should be utilised at the *nearest* available spot. If the refuse of London had been treated in this way, and had been carried to all the points of the compass for utilisation instead of being collected at one spot, it would probably have proved to be less of an incubus than it is at present.

This is the age of centralisation, of co-operation and big schemes; but in the matter of refuse disposal big schemes have not proved successful, for the very obvious reason that the greater the distance refuse is transported the greater the loss. Refuse is valuable if used on the spot and immediately. Storage and transport diminish the profit from refuse, and dilution destroys it altogether. Country towns would, it is tolerably certain, do well to carry their sewage to many points rather than one, and by so doing they would simplify their difficulties, and would be able to make use of the natural formation of the surface. It is tolerably clear that all sanitary authorities should cultivate the soil. Some have embarked in big schemes of sewage farming, but the success of these schemes has often borne an inverse proportion to their size. Every town should be planted, and perhaps some day our country towns will be beautified by sanitary authorities instead of being simply disfigured. If the streets and roadways were planted, and if the trees thus planted were nourished with some of the refuse of the town, our country towns would gain in appearance and healthiness. At least, the putting of refuse matter to its proper use might convey a valuable lesson to the inhabitants.

It would almost seem that living is healthy in proportion to the simplicity and economy which are observed. Almost all expenditure on needless luxuries in the house involves some risk; and in order to illustrate this point it will be well to glance at the modern dwelling fitted with what are termed 'the latest sanitary improvements.'

In that excellent work, entitled 'Our Homes and How to Make them Healthy,' published by Messrs. Cassell & Co., and edited by Mr. Shirley Murphy, the medical officer of the London County Council, will be found a series of articles by eminent sanitarians which are well worthy of study.

Mr. Thomas Eccleston Gibb, who writes on the legal liabilities of householders, warns us that 'there is no part of a house where builders are so likely to "scamp" their work as the drainage, and they are probably in that point under the least amount of supervision by public authorities. A local surveyor may order the work to be done in a particular manner; but the work is done and covered up in his absence, and the nicely-worded bye-laws which hang in the builder's office are not more likely to be looked at by the builder than are the drains, when buried beneath the ground, by the officer appointed to see those bye-laws carried out. Nothing short of an alteration in the system can remedy this great sanitary defect.'

The late Mr. Eassie, C.E., in a very able article, tells us of some of the difficulties and dangers of house drainage, and warns us:—

1. That the pipes sent by the maker are often so bad that the unloading of them at the railway station must be superintended by someone who represents the interests of the purchaser, and that the individual pipes must be inspected, and the bad ones rejected and returned to the manufacturer at his expense.

2. That pipes are often badly fired, too brittle, too rough on the inside, too thin and ill-fitting at the sockets.

3. That the laying of pipes is no easy matter. The ground may get sodden and sink away beneath, and thus the levels may get wrong and stoppages occur. Again, workmen will often maintain a level by means of wooden wedges, and these in time rot, and the proper level is lost.

4. That the joining of pipes involves great care. The sockets must fit and the cement be good; and we are warned that, if the cement projects into the interior of the pipe, the flow through the pipe is impeded and dangerous stoppages may occur. Again, it is common

to make a junction between a big pipe and a small one without a proper diminishing pipe, and then leakage is sure to occur, and the earth gets sodden.

5. That if pipes be laid (as often is the case) before the heavy building work is finished, they run a great risk of being broken by the falling of heavy bodies on the earth above them.

6. That pipes are so liable to become disarranged in some way that it is never safe to have them beneath a house. When they pass through the wall of a house they are very liable to break when the house settles; stoppages will occur, and the drainage will be penned back and become a source of danger.

[N.B. Houses are very liable to settle when they are built, as often is the case, on heaps of rubbish and rotting refuse.]

7. That pipes are very liable to get choked, the chief causes being (a) defects of manufacture or workmanship as indicated above; (b) collections of sediment where the level becomes deranged; (c) collections in syphon bends, which, we are informed, are 'the best-abused article in a line of drainage;' (d) the congealing of fat from the kitchen; (e) the invasion of pipes by the roots of trees—for trees have a nasty trick of driving their roots where they can get nourishment.

8. That 'grease traps' are necessary, especially when the scullery sink is at the farthest point from the sewer, as it is in the majority of London houses. These grease traps must be cleaned every two or three months, because they generate, during decomposition, 'very disagreeable smells.'

9. That having laid our drains, our next efforts must be directed to keeping back the foul air which will accumulate in a foul place; and 'traps' are necessary,

Their name is legion, and we are warned that the commonest of all (the bell-trap) is 'most reprehensible.'

10. That soil-pipes are too often made of badly socketed earthenware pipes, and when tested are not unfrequently found to leak at every joint and to be broken at the foot.

11. That as regards the proper material for soil-pipes, (a) zinc is too weak and must not be used; (b) cast-iron is not a bad material, but 'the vilest attempts at making a joint are perpetrated;' (c) lead is the best material, but then the lead must be thick enough and the jointing perfect.[1]

12. That the joints of soil-pipes must not be 'slip-joints, where one end of a pipe is slipped into the end of another and the space filled up with *no matter what!*'

13. That the soil-pipes must be properly fastened to the walls. We are told how, out of ten stacks of soil-pipes examined in a northern hospital, there was scarcely a sound joint found from want of this precaution!

14. That the soil-pipe (if it can be made air-tight) needs artificial ventilation, and the ventilator, of the same diameter as the soil-pipe, must be carried to a height of at least four feet above the highest window, and we must take care lest the birds come and build in it.

15. That 'it is constantly a matter of surprise and disgust to notice, especially in houses built for the working and middle classes, how often rainwater pipes are made to do duty for soil-pipes as well; how often the waste-pipes of baths and sinks are taken into such combined pipes; how always these pipes communicate at

[1] During a debate on the sanitary defects in houses which took place at the Parkes Museum, a story was told of an enterprising American who made a soil-pipe of old meat tins which he soldered together. The jointing in this case was sadly defective.

the foot with the house-drain, not disconnected from the sewer in any way; and how, very frequently, such pipes, doing double duty, terminate level with the top windows, giving off vaporous effluvia into the warmer room whenever the top sash is pushed down or the bottom one lifted.'

16. That it is necessary to disconnect the soil-pipe and the house-drain generally from the sewer, in order that the air of the sewer may be kept from the house, and a current of air circulate in the house-pipes.[1] The means of disconnection require an expert for the proper understanding of them, and we are warned that, 'Far too frequently, after an ordinary builder has produced what he terms "disconnection and ventilation on modern lines," the bulk of the work has to be rearranged at considerable expense.'

17. That pipes and drains so frequently being stopped, it is advisable for the owner of a large house to keep on the premises proper cleansing rods and forcing gear.

Mr. Eassie having given a number of warnings as to what *may* occur in consequence of ignorance, bad workmanship, or accident, Dr. Corfield takes up the running, and from the store of his ample experience tells us what *has* occurred. Dr. Corfield tells us:—

18. How rats make runs from the drains beneath our houses and invade the house, and how they will run from a defective drain in one house, beneath a party-wall, and up the drain of the next house.

19. How he has found cesspools leaking into wells.

20. How drains have been found with an insufficient fall, or even sloping the wrong way.

---

[1] It is evident, however, that the free circulation of air in the house-pipes must, especially in hot weather, favour the evaporation of water in syphon bends and lead to the unsealing of traps. (G. V. P.)

21. How drains have been found not jointed at all, or jointed the wrong way, so that they must leak.

22. How it has been attempted to take a drain round a corner by means of two straight pipes, meeting at an angle, instead of a properly curved pipe, and how the open angle necessarily leaked.

23. How junctions are made by means of clumsy holes roughly knocked in a big pipe, in order to take the end of a small one, and how blocks and leakage are thereby brought about.

24. How the sewer air comes up the kitchen sink.

25. How rainwater pipes, and even special ventilating pipes, bring sewer air to the attic windows.

26. How the *upper end* of a soil-pipe has been allowed to remain open and terminate *inside the house*.

27. How the *lower end* has been found to have no connection whatever with the drain.

28. How soil-pipes have been found traversing the wall of a larder, and how they have been perforated by the hooks and nails on which the mutton is hung.

29. How traps are often a delusion and a snare, and how, in addition to the bell-trap, the next commonest, the D trap, is worse than useless.

30. How lead quickly wears out and gets perforated.[1]

---

[1] In the Parkes Museum is an interesting collection of old lead sanitary fittings, which have been removed from houses, and which have been the cause of more or less illness in consequence of their becoming perforated, and thereby admitting sewer air to the house. Water-traps are of only partial use in keeping foul gases out of houses. The foul gases of the sewer are absorbed by the water on one side and given off on the other, and if there be foul gas in the sewer, the water of the trap is sure to become charged with gas quite independent of differences of pressure or temperature. If the water in the trap evaporates (as is sure to happen in any sink or closet which is little used, and is in some forgotten corner), or if the water is sucked out by the flow of water past it from some pipe above, then sewer air has, of course,

31. How sewer air may come up the waste-pipe, or over flow-pipe of the cistern, and contaminate the drinking water.

32. How 'syphon bends' get emptied by suction or choked by deposit.

33. How pan closets and D traps get plugged and 'the container' foul.¹

34. How valve-closets, solid plug-closets, wash-out closets, and Hopper-closets are each liable to their special and peculiar defects.

35. How special cisterns are necessary for the service of the w.c., and how outbreaks of typhoid have arisen from neglect of this precaution.

36. Finally, we are told how 'foul air often travels about houses by most unexpected channels. Rat-runs have already been mentioned; but besides these it travels under floors, behind panelling and wainscoting, along ventilating shafts, through defective flues, and even through the tubes in which bell wires are carried, through which foul smells from the basement, and still more frequently, the products of the combustion of gas-burners, often ascend into rooms upstairs.'

The above catalogue of common dangers in modern houses seems to show that, as all 'modern sanitary improvements' are liable to wear and tear, the danger arising from them (even assuming that they are all, to

---

free access through an untrapped pipe. We can never see the state of our 'traps,' and we can only infer (and often wrongly) that they are sealed. A distinguished physician, speaking at the Mansion House in aid of the Parkes Museum some years since, spoke of the D trap as the 'double D trap, because it deals out Death and disseminates Disease.'

¹ The writer has seen the 'safe' of a pan closet, which is intended to catch slops which accidentally spill over, perforated by a bell wire, and the spilt slops soaking along the track of the wire.

begin with, perfectly made and perfectly fitted) must be proportionate to the quantity used.

They all cost money, many of them are very expensive, and they all add to house-rent, or diminish the profits of the landlord.

When there is a small garden they are *all of them* unnecessary, and it is perfectly idle to contend that there is greater decency in the use of water apparatus than in the so-called dry methods. Dry methods do not open the door for the profit of patentees and others; but if a little of the ingenuity which has been devoted to the manufacture and subsequent exclusion of sewer air had been devoted to the easy use of dry methods of filth manipulation, many lives would have been saved and much money also.

It is the fashion of the present day to bring all that is nasty into our houses, even though those houses stand in hundreds of acres of park land. And owners will spend thousands in order that they and their households may be in constant danger of sewer gas.

If Jonathan Swift had lived on into the nineteenth century, assuredly Lemuel Gulliver would have been made to take a voyage to the modern Hygienic Laputa; and possibly Swift's wit and satire would have been able to bring people back to the straight road from which they have gone so dangerously astray.

Great cities which have got fast stuck in a sanitary quagmire must perforce pay large sums to have their troubles lessened; and to that end it is to be hoped that the almost endless clauses of Sanitary Acts will be of use. The moral duty of the individual in a city is to obey the law and assist in every way in its proper execution.

In rural and semi-rural districts, the individual ought

no more to ask others to keep him clean than he asks others to feed him or clothe him. He ought to take a pride in keeping his house wholesome and clean, and he ought to receive every encouragement from the authorities if he do so.

In connection with the question of house construction, the author would venture to make a few additional suggestions without in any degree trenching upon the domain of the architect. Architects are educated for the most part in crowded centres, where the problem of how to get the greatest amount of accommodation on the smallest area is paramount, and it too often happens that the town-bred architect, when called upon to build a house in the country, is unable to cast away the unwholesome notions which have been engrained upon him in the city, and often fails sufficiently to appreciate that the building of a country residence is a problem quite distinct from the building of a town residence, and that nothing is so conducive to the comfort, beauty, and wholesomeness of a dwelling as is an ample area upon which to construct it.

In hospital construction the rule has been for many years absolute that the sanitary offices shall be in detached turrets approached from the main structure by a short passage having thorough cross ventilation. If this be good for hospitals why is it not good for dwelling-houses? It is considered dangerous for hospital patients to live and sleep in an atmosphere polluted by sewer gas. Is it good for those who are well? In the planning of a country house, the kitchen and offices, cisterns, sinks, baths, w.c.s, &c., should be in an annexe detached, however slightly, from the main building, and approached at every floor by a lobby having cross ventilation. This annexe should contain no bedrooms, and on no account ought there to be any 'sanitary apparatus' within the four walls of the main structure. A bedroom, dressing-room, bath-room, and w.c. *en suite* may be a very nice arrangement for those who are too lazy to walk a few extra steps, but it is an arrangement which has killed its hundreds, and ought never to be employed unless the bath and w.c. be in a detached turret and approached by a lobby having cross ventilation. The greatest luxury in a house is ample area, and no country house should be more than two storeys high.

I have lately seen two country houses, recently erected, each of them standing in an ample park. In the first of these the nurseries were at the top of the house in *a third storey*, so that the children should have a maximum amount of stairs down which to fall and should inhale all the used-up air which might drift up the staircase from the floors below. It is needless to say that these nurseries had sanitary offices almost in the bedrooms. The rest of the house, which in many respects was very beautiful, showed a similar amount of thoughtlessness in the sanitary arrangement, and the owner has since died with sore throat and pneumonia.

The second of these houses was equally beautiful. A picturesque entrance hall led to a fine old English staircase which conducted to the floor above. The reception-rooms, of course, opened into the hall, which in this way, and by the staircase, provided air or might do so for every room in the house. Opening by a door *in the entrance hall*, and with no separation except the door, was a lavatory with tip-basin and w.c. ready without notice to flood the whole house with poison. The other sanitary arrangements showed an equal want of thought, and it is needless to say that such a house could never be safe.

Again, I have lately seen a school which had to be closed in consequence of a slight outbreak of diphtheria. In this school (standing in twenty acres of ground) the mistake had been made of not separating the sanitary offices from the main structure, and as the sanitary arrangements needed repair, it was necessary to delay the return of the boys until this had been done. In this school was a new building containing a dormitory, the staircase of which led direct to a big bath-room (with six or eight baths) in the basement. This arrangement ensured that if any effluvia came up the waste-pipes of the baths they might travel direct to the dormitory. In all schools, especially boys' schools, the sanitary offices, baths, lavatories, w.c.s, should be in separate structures, detached and approached by a cloister. If this were done it would be impossible to poison the boys in their bedrooms with the effluvia from drains. The accommodation in case of sudden sickness in the night should be in a detached turret. To allow boys to work or sleep under the same roof with a sewer-pipe of any kind is quite indefensible because it is perfectly unnecessary. The closing of schools in consequence of outbreaks of filth disease is getting ominously common.

School buildings should not be more than two storeys high, workrooms below, dormitories above. It is not advisable to place one dormitory over another, or one classroom over another, because this unnecessarily increases the risk of epidemics, and it must always be borne in mind that children and young people are far more susceptible of infection than adults. The example set by the London School Board is not on any account to be followed.

What is true of schools is also true of barracks. A barrack should have ample area, and two storeys should be the maximum, and the British taxpayer ought to insist that the soldiers which cost him so much are not killed or rendered useless by sewer gas.

The Royal Barracks at Dublin, so noted for typhoid, have or had three miles of sewer-pipes beneath them. A soldier at home is supposed to be fitting himself for a campaign. He ought to be systematically taught 'field sanitation,' and ought not to be allowed to poison himself with sewers. I know something of a barrack in the South of England where the field-officers' quarters are four storeys high with underground kitchens, and 'area gates' as in London. These houses are without a single square inch of private curtilage but have all 'the latest sanitary improvements,' so that it was at the time found necessary to drive a champagne cork into the overflow pipe of the bath 'to keep back the smell when the tide was rising.'

Equally important with schools and barracks are hotels, not only because the fashion of living in hotels is on the increase, but because delicate persons and invalids who are sent to so-called health resorts have to reside in hotels which are generally, from the point of view of health, of the very worst construction. When a 'health resort' becomes fashionable its doom is sealed because the value of land rises, and that crowding together of houses commences which always more than counterbalances the best hygienic efforts in other directions. The following remarks were penned after an autumn holiday, during which the writer had been 'hotel poisoned,' and are inserted here by the kind permission of the proprietors of the *Lancet*:—

It is a remarkable fact that by no means unfrequently the first idea which seizes on the mind of an hotel speculator or his architect is one which is essentially destructive. Having decided to '*exploiter*' this or that district by means of an hotel, he proceeds to erect a building which is absolutely out of harmony with its surroundings, and which tends to destroy, in so far as it be possible for one building to operate in that direction, those very characteristics which have made the locality famous or attractive. Instances will readily occur to the mind of localities which hotel-keepers have done their best to spoil. Notable examples are to be found at Arcachon, where the brute mass of a huge hotel towers above the surrounding villas and pine-trees; at Meiringen and Thun, in Switzerland, where, alongside of rows of châlets and quaint mediæval towers, the barren-minded architects have placed buildings of which the best that can be said is that they might pass unnoticed in a fifth-rate Parisian boulevard; and at Augsburg, in Bavaria, where the 'Drei Mohren,' a hostelry associated with Charles V. and the Fuggers, has been pulled down and re-erected after a model which is only too common. The first thing which the writer would impress upon hotel architects is to be careful to let the buildings harmonise with the locality. The hotel at which a traveller first alights on visiting a new district ought to serve as an introduction as it were, and prepare his mind for the pleasures which may be in store for him.

The most common fault in hotels is the fact that the buildings are too high, and occupy an area which is absolutely inadequate for the purpose. In big cities this may be a necessity, but no such necessity exists in country places, and the main reason why country hotels are built on the model of town hotels is probably to be found in the very simple fact that, as I have hinted, most architects are town-bred. The provision of adequate area for the inmates is insisted upon by designers

THE HOUSE

of hospitals, prisons, and similar institutions, and perhaps some day the same wise ideas will prevail with designers of hotels, which are institutions in which the inmates are ready to pay handsomely to be made really comfortable and to be maintained wholesomely. In hospitals an acre of ground is considered desirable for each hundred patients; and although this allowance may be excessive, it is well to bear in mind that, when numbers are congregated under one roof, it is absolutely necessary that area in proportion to the numbers be provided, and it is most important also to remember that even an excess of cubic space will not compensate for lack of area. Nearly all hotels stand upon a very deficient area; and almost all hotels when full present a degree of overcrowding which, from a sanitary point of view, is scandalous. Huge towering buildings, with as many floors as a warehouse, and perforated with staircases and hoists, so that each floor is 'ventilated' into those above and below, may be tolerated for a short time, but as places for prolonged sojourn or anything like permanent residence they are most undesirable. A resident in an hotel ought to be able to enjoy at pleasure either society or seclusion. In the public rooms he should find all that movement and gaiety which constitute the sole pabulum of dull minds, and which all of us enjoy at times; while, in his own apartment, he ought to find quiet and peacefulness in the highest degree. This end cannot be attained without adequate area, and it is notorious that one of the greatest defects of modern hotels is the impossibility of being quiet until all the inmates are in bed and asleep. In an hotel where prompt service conduces so much to comfort, it is needless to say that the sources of supply should be equally accessible from all the living rooms.

How to combine the possibility of seclusion with accessibility to the basis of supply is the problem for hotel architects to solve. It is probable that the best plan would be to place the offices and society rooms in a central structure, with the kitchen on the top floor, while the private rooms and bedrooms should be arranged in wings radiating from such central structures. The smell of the kitchen is a terrible fact in most hotels, and it is certainly surprising to find how infinitely rarely do architects take any trouble even to minimise it. I can recall one hotel, and only one (at Royat, in Auvergne), where the kitchen is placed on the top floor of the building; but I could name several in which the effluvia of the kitchen are conducted with infinite care into the bedroom corridors. The old-fashioned plan, which was common in England, of building inns round an open court and with open corridors, was good, and the fact that the court was really open probably prevented unwholesome effluvia from reaching the bedrooms. In the present day it is common to find the rooms arranged, not round an open court, but around the well of a staircase which is closed, and instead of two tiers of rooms, there are seven, eight, or even more. No worse design from the point of view of safety, comfort, or wholesomeness can well be conceived. With such a design there is no escape from noise or effluvia; there is a minimum amount of privacy, a maximum amount of labour in travelling from the front door to the rooms, and when these buildings catch fire we can only say: 'God help those who live on the top floors!' And yet this is the commonest form of hotel, and the idea seems prevalent that the height of an hotel is nowadays immaterial, because the hydraulic lift abolishes (for the guests) the labour of going upstairs. That these lofty hotels enable proprietors to increase their accommodation without increase of ground-rent there is no doubt, but there is equally no doubt that, from the guest's point of view, they offer no advantages whatever, unless (which is very doubtful) they cheapen the cost of rooms. Of the hygienic disadvantages I have already spoken, and it remains only to be said that, in order to check the drifting of effluvia and noise from floor to floor, it is absolutely essential that staircases and lifts should be placed in buildings projecting from the main building and provided with independent ventilation. The time and labour which are involved by living at or near the top of a lofty building is self-evident. Theoretically, the lift reduces the expenditure of time and labour to a minimum, but practically the guest is often willing to incur the labour of mounting rather than wait for the lift. The cost of working these lifts is, as we have seen, very great.

I have incidentally mentioned the effluvia of an hotel. These are chiefly due to humanity, cooking, tobacco, and drains, and in an average hotel the admixture of these four ingredients is, I should imagine, very tolerably equal. It is not necessary to dwell at length on the so-called 'sanitary arrangements' of hotels, especially of continental hotels. These arrangements are, as a rule, execrable; the closets are often of the worst type, the supply of water is generally deficient, and the sanitary offices practically always ventilate into the corridors or shafts of the building. In designing an hotel, provision ought to be made for placing closets and sinks in turrets approached by properly ventilated lobbies, so that the impregnation of the main building with sewer air may be a little less certain than it is at present. In big country hotels there ought to be (as there is in every country house) proper and ample accommodation provided in the gardens (approachable by a covered corridor if necessary). In this way the strain put upon the internal accommodation would be lessened, and the indecencies, often experienced in a big Swiss kurhaus, would be avoided. Certainly a gentleman's 'garden' ought to be provided as one of the sanitary accessories of an hotel, and it is probable that, properly managed, a similar arrangement would be welcomed by the ladies also. It is inconceivable to what extent this blind and dangerous practice of keeping 'sanitary offices' within the walls of the main building is carried. Quite recently I visited one of the best known show places in Switzerland—to which a railway brings hundreds of tourists daily during the season. Arrived at our destination, luncheon was necessary, and for luncheon a visit was paid to one out of several big hotels. This hotel, built in a place where the price of land can hardly be excessive, was of the usual type, four or five storeys high, with a central staircase and shaft running from the entrance hall to the roof. The entrance hall simply reeked of paraffin and ammonia *plus* hotel. The paraffin came from the lamps, the ammonia from the closets, which opened directly into the central hall. These closets communicated direct, without traps, with cesspools which received the slops and excrement of this huge establishment. To step from the pure air of the mountain top to the fearful atmosphere of this hotel was calculated to impress the mind of one who, having travelled to a spot where the air should have been absolutely pure, found that he was to run no small risk of being poisoned while he ate his lunch! At luncheon one could not but remark the anæmic condition of several of the waitresses, and this was probably accounted for by the insanitary condition of the house.

In the present day, hotel advertisements generally

THE HOUSE

lay stress on the fact that there are 'lifts,' a monosyllable which lets in a flood of light on the general plan of the building, and indicates with tolerable certainty that the area of the building is too small for the population it is designed to accommodate. Possibly some day we may have hotel advertisements couched in some such terms as these : ' This establishment, standing in extensive pleasure-grounds, has been designed on the model of a luxurious country house. The main building is only two storeys high, and has been so designed that all the bedrooms may be easily reached without any undue effort. The sanitary offices are detached from the main building. The public rooms are also detached, but they are central, equally accessible from all parts of the house, and so arranged that the smell of the kitchen cannot possibly prove an annoyance.' It is needless to say that every country hotel ought to stand in sufficient ground to enable it to be independent of any public sewer, and to deal with its own refuse in a manner which shall ensure the safety of the guests, render the soil increasingly prolific, and prove a source of profit to the landlord, rather than a perennial expense and nuisance.

Leaving the question of general design, we may consider some of the details of hotels. Thick party-walls and double doors are absolutely essential to ensure the quiet of the guests. The bedrooms, provided with these requisites, should be of a convenient form. On the Continent the hotel bedroom is generally too long and too narrow. The window is often too small, and so heavily and ridiculously upholstered as to reduce its light-giving power to a minimum. This arrangement ensures that the end of the room farthest from the window is nearly always dark, and as the looking-glass is placed against one of the side walls, the difficulties of the toilet are very great. The bedrooms are often needlessly high, and not unfrequently there is a space of two feet or more between the ceiling and the top of the window. In 'summer hotels' the bedrooms have no chimneys, so that it is impossible to get a current of air through them without opening the door, and the opening of the door generally means the admission of suffocative vapours from the corridor. Every hotel bedroom should have a covered balcony so arranged that in winter it can be enclosed by outside windows. A balcony is in summer a very pleasant addition to a room, but as an adjunct to an hotel bedroom it is necessary, in order to afford the occupant a place for shaking and brushing dusty clothing—a process constantly necessary for a traveller, but one which cannot be performed except in the open air. It is as well to state that the balcony is more common in continental than in English hotels. The thoroughly well-designed hotel bedroom, carefully 'fitted' rather than furnished with an eye to the comfort and well-being of the occupant, is one of those things which may be found when upholsterers do not happen to be the chief shareholders in hotel companies. In one particular, hotels on the Continent are uniformly first-rate, and that is in the excellence of the beds. These are always comfortable, and very superior to the beds which are found in English hotels. Of the rest of the furniture, the less said the better; but I cannot refrain from commenting on the perverted genius of the man who invented the chest of drawers and washstand in one, so designed that one's clean shirts are almost inevitably invaded by a flood of soapsuds. In a tour of twenty-eight days, 75c. per diem, or a total of 21 fr., was paid by the writer for the tin vessel, and some two gallons of water, which the Englishman (especially when engaged in active exercise) has the arrogance to demand for the purpose of cleansing his skin. One need not grumble at the sum paid, but certainly one has a right to grumble at the difficulty often experienced in getting this customary sitz-bath. Perhaps, when the Germans have written a few more treatises on 'Haut Cultur,' hotel proprietors will provide as a matter of course that which is now wrung from them with difficulty. The writer can only recall one hotel in which he has found a sitz-bath supplied without asking, and that hotel was at Malta, and it is probable that the landlord had been drilled in this particular by officers of the English garrison.

CHAPTER VI

AIR

WE all know the importance of fresh air. Our instincts have told us this from all time. Nowadays every child is taught in the Board Schools the scientific proofs of why we need fresh air, so that any lengthy disquisition of this point will not be necessary.

We breathe some sixteen times in a minute, and we take in nearly a pint of air at each breath, or two gallons every minute, or 120 gallons every hour, or 2,880 gallons every twenty-four hours. Each pint of *fresh* air contains about 15·8 fluid ounces of nitrogen, with 4·19 fluid ounces of oxygen, and to this is added ·008 fluid ounce of carbonic acid. We use up the oxygen and we give off carbonic acid, so that every pint of expired breath contains still 15·8 fluid ounces of nitrogen, with 3·26 fluid ounces of oxygen, and 0·94 fluid ounce of carbonic acid. Expired air contains in addition much watery vapour together with organic matters, which we recognise in 'the smell of humanity,' which is always present in a crowded room, or a closed bedroom which has been slept in.

Air which has been breathed once is poisonous, and a man in an hermetically sealed room would soon die.

The constant admission of fresh air to rooms is an absolute necessity, and were it not for chimneys, ill-fitting doors and windows, keyholes, chinks and crannies in the walls, and the accidental opening and shutting of doors, suffocation would be a far more common mode of death than it is.

In order that the air of an apartment may be kept wholesome, our breath needs constant dilution with fresh air; and if we are to keep the amount of carbonic acid anywhere near its normal point, it is obvious that about 100 pints of fresh air must be admitted for each breath which is drawn by every person in the room, or about 12,000 gallons per head per hour.

For every person in an apartment, there ought to be an air-hole having the diameter of a gallon measure, and through this the air should move, with a velocity equal to 12,000 times the height of a gallon measure, or about 7,000 feet per hour, or about 120 feet per minute, or 2 feet per second. For every inlet there must be an outlet of equal size. If the apartment be small and the air-inlet in a bad position, draughts will be created, because the large proportion borne by the incoming air to the total air in the apartment will cause currents to be felt everywhere. If the apartment be big and the inlet be $6\frac{1}{2}$ or 7 feet from the ground, and deliver its air vertically, and if (in cold weather) the incoming air be warmed, no draught will be caused.

As regards outlets, the ordinary open fireplace is usually sufficient. It is not advisable to give less than 1,000 cubic feet of space to each occupant of a room. In prisons, each cell contains about 800 cubic feet of space, and, says Professor de Chaumont, 'practically this is found to be too small.'

Each gas-flame or lamp uses up the air just as a human being does, so that, in calculating the amount of cubic space necessary, we must reckon each lamp or gas-burner as an individual. It must be remembered that great cubic space is of no use unless inlets for fresh and outlets for foul air be also provided.

Ordinary churches are, as a rule, very badly ventilated. The cubic space is enormous, but its quantity is due entirely to height. The area in proportion to the congregation is very small indeed, and the 700 or 800 people in an ordinary church almost touch each other. In addition to the people there are often a great many gas-burners. There is often no ventilation, and it is a growing custom to replace the plain window, which might be opened, but seldom was, by a painted window which cannot be opened even if desired. What is the consequence of all this? The people give off their 120 gallons per head per hour of hot foul breath; this ascends towards the roof of the nave, and being cooled, sinks (for carbonic acid is heavy) and envelopes the congregation, as it were, with a soporific pall. Some faint, others go to sleep, and the preacher, poor man! perhaps fancies that the sermon, and not the carbonic acid, is the narcotic which has acted on his flock.

If any aspirant for clerical honours wishes to gain a reputation as a forcible and enlivening preacher, let him first ventilate his church, and let him be sure, whenever he puts up a stained window which will not open, to compensate for the loss by putting in a ventilator in some other place.

Let him also remember that one of the liveliest and most successful preachers that ever adorned our Church, honest Hugh Latimer, gained his reputation while preaching in *the open air* at Paul's Cross. In like manner the theatres, which so successfully spurred the intellects of the greatest poets of the world (the Greek dramatists and Shakespeare), were freely exposed to the fresh air, and poet and actor alike had the advantage of audiences in which the critical faculty was neither blunted nor savaged by the atmospheric foulness of the place.

We live in an age of public meetings, and throughout the country there are hundreds or perhaps thousands of gatherings every day for the purpose of discussing questions of public interest. How few of these meeting-rooms are adequately ventilated ; and how much harm is done by sitting for hours and breathing your neighbour's breath almost undiluted, it would be difficult to say! It is tolerably certain that 'colds' are caught by sitting in foul rooms. If air is deficient in oxygen and is loaded with organic vapours, the elimination of refuse matter from the blood does not go on properly, and when we have reduced our bodies to a state to be affected by any untoward circumstance we suddenly chill the surface by opening a window or going into the cold night air, and then we blame the latter circumstance only, and give little attention to the two or three hours' preparation for mischief which we had previously undergone.

The *open air*, even in the most crowded London streets, is always infinitely more pure than the air of even well-ventilated rooms. The reason for this is that the volume of our atmosphere is, as compared with the volume of foul air which escapes from our houses, almost infinite, and the dilution which foul air undergoes is infinite.

The foul air is lost as soon as it escapes from our houses. It is diffused, mixed and blown away. Air moves at the average rate of 10 miles per hour, 17,600 yards, or 52,800 feet. Taking the area occupied by a man at 9 square feet, we find that the astonishing quantity of 475,200 cubic feet, or 2,980,000 gallons, of

L

air per hour rush over the surface of the body of a man exposed to the open air. It must be remembered also that in streets and narrow channels the rate at which the air travels is often greater than in open places, and that in times of storm and wind the rate at which the air travels may be four or six times the average. Thus we see that the air which blows over one man in the open is enough to meet the respiratory needs of 1,000. If we take the average London street, 50 feet wide, and flanked by houses 50 feet high, the area of the cross-section of such a street would be 2,500 square feet, and the amount of air passing through each part of such a street per hour would amount to an average of 132,000,000 cubic feet, or 825,000,000 gallons, or more than enough for 68,000 men. Thus we see that the purifying action of the air is, by its enormous volume and diluting power, practically infinite. The amount of fouling of the atmosphere by the whole animal life of the world is but as the most microscopic drop in the bucket. We now see why it is that infection rarely travels for any distance through the open air. Well-established cases of infection being blown from house to house (in the absence of any subterranean communication by sewers, community of water-supply, or personal intercourse by laundresses, tradesmen or others) are almost unknown. This is explained by the above facts, as well as by the further fact that most organic poisons are quickly oxidised and destroyed by the oxygen and ozone in the air.

' A consideration of the above facts throws no little doubt upon the teaching of some of our hospital architects, who insist upon the necessity of separating the various pavilions containing the sick by enormous interspaces. As long as the air admitted to a sick-room comes really *from the outside* that air will be practically pure, and it will make no appreciable difference whether the next pavilion is 50 or 1,000 feet away. We must be sure, however, that the air employed for ventilation is really outside air, and that it has not been used before in kitchens, or more noisome places, and is not merely allowed to drift from one part of a building to another, whether by staircases, corridors, lifts or shoots.

The air which has been fouled by the respiration of men and animals, by the combustion of gas, oils and fuel, and by the exhalations from filth, is not only being constantly diluted but it is constantly being purified. What we give off, plants need; what plants give off, we need. The carbonic acid which escapes from our breath is absorbed by the green leaves of plants, which convert the carbon into starch and allied bodies, and give back a great part of the oxygen in a pure state. During night, it is true, plants give off carbonic acid, but it is equally true that at night the respiratory needs of men and animals are at a minimum. The influence of light, therefore, by stimulating the elimination of oxygen by green leaves, has great power over the condition of the atmosphere. Analyses of air show that there is less carbonic acid in the neighbourhood of luxuriant vegetation than where vegetation is absent. This difference is not great, because of the very free movement of the atmosphere, but it is quite enough to be measurable. In close courts of cities the carbonic acid is notably increased; and in crowded places, such as schoolrooms and theatres, the amount is dangerously great.

Although rare, it is well to bear in mind that infection can and does travel through the open air from house to house. This was well illustrated in the classic instance of the spread of small-pox from the Fulham Smallpox Hospital.

In the supplement to the Tenth Report of the Local Government Board for 1880-81 is a report by Mr. W. H. Power on the influence of the Fulham Hospital (for small-pox) on the neighbourhood surrounding it. Mr. Power investigated the incidence of small-pox on the neighbourhood, both before and after the establishment of the hospital. He found that, in the year included between March 1876 and March 1877, before the establishment of the hospital, the incidence of small-pox on houses in Chelsea, Fulham, and Kensington amounted to ·41 per cent. (*i.e.* that one house out of every 244 was attacked by small-pox in the ordinary way), and that the area enclosed by a circle having a radius of one mile round the spot where the hospital was subsequently established (called in the report the 'special area') was, as a matter of fact, rather more free from small-pox than the rest of the district. After the establishment of the hospital in March 1877, the amount of small-pox in the 'special area' round the hospital very notably increased, as is shown in the table by Mr. Power, given on p. 149.

The table shows conclusively that the houses nearest the hospital were in the greatest danger of small-pox. It might naturally be supposed that the excessive incidence of the disease upon the houses nearest to the hospital was due to business traffic between the hospital and the dwellers in the neighbourhood, and Mr. Power admits that he started on his investigation with this belief, but with the prosecution of his work he found such a theory untenable.

Now the source of infection in cases of small-pox is often more easy to find than in cases of some other forms of infectious disease, and mainly for two reasons :—

1. That the onset of small-pox is usually sudden and striking, such as is not likely to escape observation.

2. That the so-called incubative period is very definite and regular, being just a fortnight from infection to eruption.

The old experiments of inoculation practised on our forefathers have taught us that from inoculation to the first appearance of the rash is just twelve days. Given a case of small-pox then, one has only to go carefully over the doings and movements of the patient on the days about a fortnight preceding in order to succeed very often in finding the source of infection.

In the fortnight ending February 5, 1882, forty-one houses were attacked by small-pox in the special mile circle round the hospital, and in this limited outbreak it

ADMISSIONS OF ACUTE SMALL-POX TO FULHAM HOSPITAL, AND INCIDENCE OF SMALL-POX UPON HOUSES IN SEVERAL DIVISIONS OF THE SPECIAL AREA, DURING FIVE EPIDEMIC PERIODS.

Cases of acute small-pox	The epidemic periods since opening of hospital	Incidence on every 100 Houses within Special Area and its Divisions				
		On total special area	On small circle, 0–¼ mile	On first ring, ¼–½ mile	On second ring, ½–¾ mile	On third ring, ¾–1 mile
327	March–December 1877 .	1·10	3·47	1·37	1·27	·36
714	January–September 1878	1·80	4·62	2·55	1·84	·67
679	September 1878–October 1879	1·68	4·40	2·63	1·49	·64
292	October 1879–December 1880.	·58	1·85	1·06	·30	·28
515	December 1880–April 1881. . . . . . , .	1·21	2·00	1·54	1·25	·61
2,527	Five periods	6·37	16·34	9·15	6·15	2·56

was found, as previously, that the severity of incidence bore an exact inverse proportion to the distance from the hospital.

The greater part of these were attacked in the five

days, January 26-30, 1881, and in seeking for the source of infection of these cases special attention was directed to the time, about a fortnight previous, viz. January 12-17, 1881. The comings and goings of all who had been directly connected with the hospital (ambulances, visitors, patients, staff, nurses, &c.) were especially inquired into, but with almost negative result, and Mr. Power was reluctantly forced to the conclusion that small-pox poison had been disseminated through the air.

During the period when the infection did spread, the atmospheric conditions were such as would be likely to favour the dissemination of particulate matter. Mr. Power says: 'Familiar illustration of that conveyance of particulate matter, which I am here including in the term "dissemination," is seen, summer and winter, in the movements of particles forming mist and fog. The chief of these are, of course, water particles; but these carry gently about with them, in an unaltered form, other matters that have been suspended in the atmosphere, and these other matters, during the almost absolute stillness attending the formation of dew and hoar frost, sink earthwards, and may often be recognised after their deposit. As to the capacity of fogs to this end, no Londoner needs instruction; and few persons can have failed to notice the immense distances that odours will travel on the "air-breaths" of a still summer night. And there are reasons which require us to believe particulate matter to be more easy of suspension in an unchanged form during any remarkable calmness of atmosphere. Even quite conspicuous objects, such as cobwebs, may be held up in the air under such conditions. Probably there are few observant persons of rural habits who cannot call to mind one or another still autumn morning, when from a cloudless, though perhaps hazy, sky, they have noted, over a wide area, steady descent of countless spider-webs, many of them well-nigh perfect in all details of their construction.'

A reference to the meteorological returns issued by the Registrar-General shows that on January 12, 1881, began a period of severe frost, characterised by still, sometimes foggy, weather, with occasional light airs from nearly all points of the compass. This state of affairs continued till January 18, when there was a notable snowstorm and a gale from the E.N.E. For four days, up to and inclusive of January 8, ozone was present in more than its usual amounts. During January 9–16 it was absent. On January 17 it reappeared, and on January 18 it was abundant. Similar meteorological conditions (calm, and no ozone) were found to precede previous epidemics.

Mr. Power's report, with regard to Fulham, seems conclusive, and there is a strong impression that hospitals other than Fulham have served as centres of dissemination.

M. Bertillon, of Paris, quoted figures in support of that opinion. It is a fact of some importance to remember that small-pox is one of those diseases which has a peculiar odour, recognisable by the expert. As to its conveyance for long distances through the air, there are some curious facts quoted by Professor Waterhouse, of Cambridge, Massachusetts, in a letter addressed to Dr. Haygarth at the close of the last century. Professor Waterhouse states that at Boston there was a small-pox hospital on one side of a river, and opposite it, 1,500 yards away, was a dockyard, where, on a certain misty, foggy day, with light airs just moving in a direction from the hospital to the dockyard, ten men were working. Twelve days later all but two of these men were down with small-pox, and the only possible source of infection was the hospital across the river.

There are certain poisons which seem to travel mainly through the air. These are influenza, measles, whooping-cough, small-pox, typhus fever, scarlet fever, diphtheria, mumps, and chicken-pox. Typhoid fever may also travel through the air, and recent investigations have shown that that commonest of common diseases, consumption or phthisis, is distinctly communicable through the air. We have seen that contagion in the open air is so diffused that it is more likely than not to be harmless, even supposing that it be not destroyed. In dwellings the risk of contagion through the air is very great, and a case of measles or whooping-cough, even though it be confined to one room in a house, is very likely to infect any other children who may be in the house. In overcrowded and ill-ventilated rooms, contagious particles which are given off have but little chance of escape, and are very likely to be inhaled by somebody else. Crowded gatherings of children must be reckoned among the great causes of the dissemination of many of the diseases of childhood. The risks of contagion through the air are diminished in proportion to the thoroughness of the ventilation.

It is an undeniable fact also that consumption of the lungs has been in many places greatly diminished by proper ventilation and also by proper drainage, thereby causing greater healthiness and dryness of the dwelling, and getting rid of a sewage-sodden soil.

There can be no doubt that our moral responsibilities with regard to the air we breathe are very great. Our first duty is not to foul the air more than we can help, to keep all about us clean and pure, and not to allow heaps of evil-smelling refuse to collect about our dwellings.

Our next duty is to see that a proper supply of fresh air is admitted to our dwellings. If this be done there will be a higher standard of health in the dwelling, more food will be needed, more work will be done. Of the economy of giving an ample supply of fresh air there can be no doubt. Employers of labour should remember this, and especially those who employ young people. All work-rooms should have ample cubic space and free admission of air. The master will then have more work done, and more cheerfully performed, than otherwise would be the case.

Take care that every gas-light is provided with a flue communicating with the outside air. There is no real difficulty in accomplishing this; and if it be done, not only will the gas-light not *foul* the air, but it will aid in the ventilation of the room just as the fire does, by creating a draught up its little chimney.

In school-rooms and other places where mental work is to be done, good ventilation is of the greatest importance, and has a very great influence on the quality of the work done.

We must all of us try to set a good example in this matter of cubic space and ventilation. When, for example, we wish to show hospitality to our friends, we must remember not to stint the supply of the prime necessary of life. The average London dining-room is perhaps 20 feet by 16 feet by 12, and contains, inclusive of the space occupied by furniture, &c., less than 4,000 cubic feet, or space considered sufficient for five convicts in prison. If we wish to do honour to our guests we invite sometimes as many as eighteen, and to wait upon them we employ four servants, and we light the room with half-a-dozen lamps or their equivalent—*i.e.* we put into our 4,000 cubic feet of space the equivalent of twenty-eight people, and we give them 143 cubic feet of space each, and as we provide no adequate inlet or outlet for fresh air, it is not to be wondered at that the discomfort often reaches agony point, and that the conversation lags; nor is it a matter of surprise that the average London dinner, where you are suffocated and over-fed, is reckoned among the duties rather than the pleasures of existence, and that the malaise of the following day is (often wrongly) attributed to the quality of the wine.

The 'At home,' where 150 persons (not reckoning the lights) crowd into about 8,000 cubic feet of space, with something like 50 cubic feet of space each, is, as is demonstrated by arithmetic, as nearly as possible three times as bad as the dinner. Perhaps the day will dawn when it will be considered 'bad form' to give your guests not more than one-twentieth of the cubic space, and far less than one-twentieth of the fresh air, which is allotted to criminals.

Again, will nobody set us an example of keeping good hours? 'Early to bed and early to rise,' says the old proverb, 'makes a man healthy and wealthy and wise.' If we keep what are known as 'bad' hours (*i.e.* the hours till lately kept by the House of Commons), we are perforce obliged to spend those hours in rooms artificially lighted and warmed, and instead of breathing fresh air we breathe foul. The evils arising from this need not be dwelt upon. Such a state of existence is hardly compatible with good health. The M.P. at the end of the session, and the young lady at the end of the season, are standing examples of this fact. During the summer there are about fifteen hours of daylight and nine of darkness. It is the fashion to rise about seven hours after sunrise, and to retire about seven hours after sunset, and the masses follow the leaders of fashion. Will anybody calculate the unnecessary waste of gas and other illuminants caused by our obstinate refusal to make use of the sunlight? How many cubic feet of carbonic acid and sulphur compounds are poured into the London air in consequence of this perversity? How much unnecessary smoke is poured into it from the same cause? What advantage, if any, is got by converting night into day?

We have dwelt mainly upon the pollution of the air by respiration and organic refuse. The pollution by inorganic matter is, in large cities, scarcely less important. The London atmosphere is the dirtiest in the world. The skin and linen of the Londoner are grimy in a few hours after cleansing. As a consequence the Londoner is always washing, and he makes a boast of his enforced cleanliness, instead of being ashamed of the grimy cause of it. It is always well to make a virtue of necessity.

The London air is loaded with soot from chimneys, and with carbonic acid and sulphur compounds, the result of the combustion of coal and gas. In dry weather also it reeks of ammonia given off from the streets unwashed by rain and covered with horse-droppings. Only a very few plants will live in London, and none of them can be said to flourish. There is not a rose and scarcely a fir-tree of any kind in the metropolis. The leaves of plants are choked with soot, and they are killed by the acid in the air.

This state of things has grown with the growth of the city. At present there are probably few ladies strong enough to walk from Charing Cross in a straight line into the real country in any direction.

If the wind were to drop to a dead calm for a week many of the dwellers in the centre of London would certainly be killed or seriously affected by the overcharged atmosphere. When cold and calm coincide we get fog and darkness. In December 1873 the death-rate in the central districts rose from 18 to 43 per 1,000 from this cause alone, and deaths from diseases of the respiratory organs were 551 above the average. At this time it was found necessary to remove 36 out of every 100 animals exhibited at the Smithfield show, and of those removed many had to be killed. They were victims to fog.

We have not improved since then, and, in spite of all our efforts, the smoke will remain proportionate to the number of houses.

There can be little doubt that the domestic fireplace is the main cause of the trouble, and therefore it becomes the moral duty of the householder to burn as few coals as possible in order to diminish the smoke. The better the combustion of the coals the greater the heat given off; less fuel is consumed, and less smoke escapes up the chimney. It is cheap and comfortable to diminish the smoke of our fires, another added to the many instances already given of the thriftiness of good sanitary morals.

Grates of fire-clay with solid bottoms are a great improvement on the old-fashioned grates, and save quite a fourth of the coal. Here is comfort and hope.

It is difficult, however, to get even small improvements effected in houses held on short lease, and especially when improvements are 'compensated' for by increased rents and rates. Our comfort dwindles and our hopes are dashed.

CHAPTER VII

WATER

WATER is an article of first necessity to all of us. Without pure water there cannot be health. Pure water has served moralists of all times as a symbol of purity. The Christian Sacrament of Baptism is an instance of this.

Our moral responsibility with regard to water should be to regard its purity as something too sacred to be defiled. In this Christian land, however, there is scarcely a watercourse which is not polluted, and many of our loveliest rivers have been wantonly converted into sewers.

In his introduction to 'The Crown of Wild Olive,' Professor Ruskin gives an eloquent description of our swinish apathy with regard to water. The passage is so beautiful, that no apology is needed for quoting it in full.

'Twenty years ago there was no lovelier piece of lowland scenery in South England, nor any more pathetic in the world, by its expression of sweet human character and life, than that immediately bordering on the source of the Wandle, and including the low moors of Addington, and the villages of Beddington and Carshalton, with all their pools and streams. No clearer or diviner waters ever sang with constant lips of the hand which "giveth rain from heaven"; no pasture ever lightened in spring-time with more passionate blossoming; no sweeter homes ever hallowed the heart of the passer-by with their pride of peaceful gladness—fain hidden, yet full confessed. The place remains (1870) nearly unchanged in its larger features; but with deliberate mind I say, that I have never seen anything so ghastly in its inner tragic meaning—not in Pisan Maremma—not by Campagna tomb—not by the sand-isles of the Torcellan shore—as the slow stealing of aspects of reckless, indolent, animal neglect over the delicate sweetness of that English scene. Nor is any blasphemy or impiety, any frantic saying or godless thought, more appalling to me, using the best power of judgment I have to discern its sense and scope, than the insolent defiling of those springs by the human herds that drink of them. Just where the welling of stainless water, trembling and pure, like a body of light, enters the pool of Carshalton, cutting itself a radiant channel down to the gravel, through ways of feathery reeds, all waving, which it traverses with its deep threads of clearness, like the Chalcedony in Moss-agate, starred here and there with the white Grenouillette; just in the very rush and murmur of the first spreading currents, the human wretches of the place cast their street and house foulness; heaps of dust and slime and broken shreds of old metal, and rags of putrid clothes, which, having neither energy to cart away, nor decency enough to dig into the ground, they thus shed into the stream, to diffuse what venom of it will float and melt, far away, in all places where God meant those waters to bring joy and health. And in a little pool behind some houses farther in the village, where another spring rises, the shattered stones of the well, and of the little fretted channel which was long ago built and traced for it by gentle hands, lie scattered, each from each, under a rugged bank of mortar and scoria, and bricklayers' refuse, on one side, which the clean water, nevertheless, chastises to purity; but it cannot conquer the dead earth beyond; and then circled and coiled under festering scum, the stagnant edge of the pool effaces itself into a slope of black slime, the accumulation of indolent years. Half-a-dozen men, with one day's work, could cleanse those pools and trim the flowers about their banks, and make every breath of summer air above them rich with cool balm, and every glittering wave medicinal, as if it ran, troubled only by angels from the porch of Bethesda. But that day's work is never given, nor, I suppose, will be; nor will any joy be possible to heart of man for evermore, about those wells of English water.'

If poetry may be defined as the art of conveying absolute truths in the most beautiful and forcible language attainable, of at once compelling the intellect and gratifying the senses, then the above passage must take a high rank among short English poems, for its beauty is equalled by its matter-of-fact truth.

Unfortunately for the purity of English waters, the Public Health Act of 1848 encouraged the emptying of town sewerage into rivers, and we are still taught as one of the chief tenets of our sanitary creed that we should dirty as much water as possible in washing away from our houses filth which ought to be buried. As a consequence of this, pure water is becoming daily more difficult to get, and nowadays it is considered safer and better to drink water—hard, charged with carbonic acid, and deficient in oxygen—which has been raised at infinite cost from the depths of the earth, than to drink of the 'brook which babbles by,' with every bubble freshened by the air and charged with its maximum amount of oxygen. The reason for this is that the brook has almost certainly been fouled by receiving the filthiness from dwellings nearer to its source, and the natural consequence is the reflection that, if the brook has already been fouled, a little more fouling can do no harm, and thus the brook gathers sewage as it flows, till, having passed through sundry towns in its course, it flows out to sea a seething, stinking sewer.

We all of us deplore this state of things, but few of us, in thinking of the cause of such filthy impurity, ever pause to put to ourselves the solemn question, 'Is it I?' We inveigh against the 'Board,' we say that such a state of things is disgraceful, we shut our eyes to the fact that the disgrace falls upon ourselves as well as others; and even though we may be favourably circumstanced for doing our duty towards the watercourses, the pangs of conscience are seldom sufficient to make us stop our quota of pollutions; at least to do our own duty, and, doing it, set a good example to others.

Richard the Second (whose advisers probably remembered the epidemics of 'the Black Death' in the reign of his grandfather) passed an Act in 1388 which imposed a penalty of twenty pounds (worth how much of our money?) on persons who fouled ditches and rivers with filth and refuse, and in 1876 Parliament passed an Act intended to save rivers from pollution. This Act is put in force against 'Boards' and 'Authorities' (with how much success the Thames, the Mersey, and the Clyde will testify), but is seldom enforced against individuals; and although it is often easy for an individual to cease polluting a watercourse, it is often impossible for a sanitary authority to do so in the face of the apathy of the individuals by whom the members of the 'Board' have been elected.

Not unfrequently the 'Board' is content to let the individuals alone, because millions spent in sewers and other millions spent in waterworks is 'good for trade' in general, and, possibly, specially good for the special trade of some of the members of our local parliaments.

The demon of self-interest has always to be reckoned with when devising measures intended to benefit the public health.

Many diseases are caused or conveyed by impure water, the most notable being cholera and enteric fever. Let us take the commonest of these diseases, enteric or typhoid fever. The patient who is attacked with typhoid is attacked insidiously; he suffers from the disease, generally days, sometimes weeks, before its nature is recognised. The poisonous excreta of this patient pass into a watercourse or perhaps into a cesspool (a pit in which excrement and water are commingled), and the water leaks from the cesspool into the well, and then those who drink of the well suffer in their turn from typhoid.

Water thus poisoned by leakage usually bears evidence, both physical and chemical, of its contamination, but this does not appear to be necessarily the case, and there are instances of tainted waters being pleasant to the eye and palate.

That the germs of cholera and typhoid will live in water is certain, and their power of diffusion through water is infinite. One dejection from a typhoid patient is theoretically capable of infecting an almost unlimited volume of water. The mischief which one case of typhoid may do is told by Professor de Chaumont in 'Our Homes.'

'A very remarkable case was investigated a few years

ago by Dr. Thorne Thorne at Caterham and Redhill, Surrey. The Caterham Water Company found that they were unable to supply the whole district with their existing arrangements, and in the more remote part of the district they were obliged to get part of their supply from a neighbouring company. In the meantime they determined to enlarge their sources of supply by digging additional wells, and cutting and enlarging the adits from one to the other. Careful arrangements were made to prevent contamination of the water during the work, and the men were instructed to carefully avoid fouling the water with any excremental matter. One of the workmen, newly taken on, was suffering, unknown to himself, from a mild attack of typhoid fever, accompanied with diarrhœa, and he confessed that he was obliged not only to have resort to the buckets, but even to make use of the adit itself, on emergency. About twelve or fourteen days after he began to work typhoid fever began to show itself among the consumers of the water; the disease spread rapidly, and about 350 cases with several deaths took place. When Dr. Thorne Thorne investigated the circumstances, one remarkable fact became evident—viz. that the disease was almost entirely confined to that part of the district supplied with the company's water pure and simple, whilst the outlying part, which was only partially supplied from the company's wells, but whose chief supply was from those of the neighbouring company, remained nearly free from disease. This fact, joined with the other, that the disease broke out just about the usual time after the workman must have been the cause of contaminating the well, pointed clearly to the Caterham Company's water as the medium of contagion. Another corroborating fact was, that at the Lunatic Asylum, where the water-supply was from a deep well on their own premises, the inmates remained free from the disease; and at the barracks, the Guards, who also drank the water of the Asylum well, did not suffer. The latter was a pure water, as I had an opportunity of analysing it myself. The remedial measures adopted were to stop the supply of water at once, to pump the wells dry several times, to scrape the sides of the wells and the adits and wash them with chloride of lime, and to throw large quantities of Condy's fluid into the water. From that time the disease entirely ceased. No more marked proof could be given of the transmission of the disease through water.'

No more marked proof could be given of the enormous diffusion which takes place when typhoid poison is mixed with water, and of the dangers which necessarily attend upon water-carried sewage. If we foul the brooks, rivers and wells which are about our houses we must rely on water companies for the first necessary of life, but if the common source gets poisoned we encounter epidemics of an extent unknown before.

The following case which is quoted (in German) in the sixth report of the Rivers Pollution Commissioners tells a similar tale; and it also tells us that typhoid poison cannot be removed from water by the most perfect underground filtration. The ensuing version of the 'Lausen case' is taken, however, from Mr. Noel Hartley's little book on 'Water, Air and Disinfectants,' published by the Society for Promoting Christian Knowledge.

In the village of Lausen, near Basle, in Switzerland, which had never within the memory of man been visited by epidemic typhoid, and in which not even a single case had occurred for many years, there broke out in August, 1882, an epidemic, which simultaneously attacked a large portion of the inhabitants. About a mile from Lausen, and separated from it by the mountainous ridge of the Stockhalden, which was probably an old moraine from the glacial epoch, lies a small parallel valley—the Fürlerthal. In an isolated farmhouse, situated in this valley, a farmer, who had just returned from a long journey, was attacked by typhoid fever on June 10. During the next two months three other cases occurred in the same house. The inhabitants of Lausen were entirely ignorant of what had occurred in this solitary mountain farm, which was cut off from all communication with the rest of the world; when on August 7 ten of the villagers were suddenly struck down by typhoid fever, whilst during the next nine days the number of cases had already increased to 57 out of a population of 780 persons living in 90 houses. In the first four weeks the number of cases reached 100 (that is to say, that out of every 100 persons in the village more than 12 were attacked); and altogether, to the close of the epidemic at the end of October, 130 persons, or 17 in every 100 of the population, were attacked, besides 14 children who were infected at Lausen during their summer holidays, and became ill after their return to school in other localities. The fever cases were pretty equally distributed throughout the entire village, but those houses (six in number) which were supplied with water from their own private wells, and not from the public fountains, were entirely exempt. This remarkable difference led to a suspicion that the public water-supply was connected with the cause of the epidemic, although the apparent immaculate source of this supply seemed to negative any such suspicion.

The water came from a spring situated at the foot of the adjacent Stockhalden ridge. It was then received in a tank lined with brickwork and carefully protected from pollution; nevertheless, a careful investigation of the source of this spring placed beyond doubt the origin of the infection.

Ten years previously it had been proved that direct water communication through the intervening mountains existed between the spring and a brook in the Fürlerthal, flowing past the farmhouse in which the typhoid fever cases occurred. At that time (*i.e.* ten years before) there was spontaneously formed, by the giving way of the soil for a short distance below the farmhouse and close to the brook, a hole about 8 feet deep and 3 feet in diameter, at the bottom of which a moderately clear stream of water was observed to be flowing. As an experiment the whole of the brook water was now diverted into this hole, at the bottom of which it entirely disappeared, but in an hour or two the spring at Lausen, at that time nearly dry from a long drought, overflowed with an abundance of water which was turbid at first, but afterwards clear, and this overflow continued until the Fürler brook was again confined to its bed. It was, however, afterwards noticed that, whenever the meadows below this hole were irrigated with the water of the Fürler brook, the volume of the Lausen water-supply became greatly augmented a few hours afterwards. Now this irrigation, practised every year, was carried on in the summer of the epidemic from the middle to the end of July, the brook being polluted by the dejections of the typhoid patients—for it was in direct communication with the closets and dung-heaps of the infected house, whilst all the chamber-slops were emptied directly into it, and the dirty linen of the patients washed in it. Soon after the irrigation had begun the water-supply to Lausen, which was at first turbid, acquired an unpleasant taste, and increased in volume. About three weeks after the commencement of the irrigation the sudden outbreak of typhoid fever in Lausen occurred.

In his search after the cause of this outbreak, Dr. Hägler, of Basle, did not rest satisfied with the evidence just recorded, but supplemented it by the following ingenious and conclusive experiments: The hole in the Fürler was reopened and the brook again led into it; three hours later the fountains at Lausen delivered double the quantity of water.

'Eighteen hundredweight of salt, previously dissolved in water, was now poured into the hole, and soon the water at Lausen exhibited a great increase of saltness, until the solid matter in the water increased threefold. The passage of the Fürlerthal water to the fountains of the fever-stricken village was thus ascertained beyond doubt. But another interesting question here presented itself: did the water find its way through the Stockhalden by a natural open conduit, or was it filtered through the porous material of the old moraine?'

'To decide this point $2\frac{1}{2}$ tons of flour were first carefully and uniformly diffused in water, and then thrown into the hole; but neither an increase in the solid constituents nor the slightest turbidity of the Lausen water was observed after this addition.'

This remarkable case shows:—

1. That the power of mischief possessed by water-carried sewage is enormous.

2. That the diffusibility of typhoid poison in water is practically infinite.

3. That water containing typhoid poison may not be purified by filtration through nearly a mile of solid earth (a filter fine enough to arrest particles of wheat flour), although it must be borne in mind that the filter in this instance was deeply buried and that the certain absence

of air and aerobic microbes rendered the oxidation and purification of the water passing through it an impossibility. The fact, also, that the Lausen water became *turbid* when the Fürler brook overflowed seems to indicate that, although the flour-experiment gave a negative result, solid particles were, nevertheless, able to find their way through the Stockhalden ridge.

4. That large typhoid epidemics are favoured by a water-supply common to many people, if by mischance that water-supply gets fouled.

Medical literature is crowded with instances of mischief caused by water being contaminated by leakage from sewers and cesspools. The fact is so well established that it is not necessary to weary the reader with instances. The above cases show clearly, (1) that one man has infected 350 others; and (2) that infections may travel for a mile through an *underground* filter.

In the face of the Lausen case it would almost seem that the absolute protection of a water-supply is nearly impossible.

Deep wells which are sunk in chalk or any other porous soil are liable to pollution from foulness finding its way into them from the surface or through cracks or fissures in the soil, and this danger is proportionate to the amount of water pumped from the well. Professor de Chaumont says[1]: 'The area of surface drained by wells is a question of some difficulty. It has been stated as a circle the radius of which is the depth of the well; but this appears to be a grave understatement of the case, if we look to the evidence which has been obtained from the effects of pumping upon distant wells, or the way in which wells have sometimes been drained by outflows of water at distant lower levels. A

[1] *Our Homes*, p. 787.

well in a gravel and sandy soil in South Hampshire was
found to be drained dry in consequence of an outflow of
water in a gravel pit dug a considerable distance off.
The difference of level between the higher point (that is,
the bottom of the well) and the lower (the outflow at the
gravel pit) was $21\frac{1}{2}$ feet, the distance between the two
1,720 feet ; so that the area drained had a radius equal
to eighty times the depth, here represented by the fall
or difference in level between the two points.'

Messrs. Rogers Field and Wallace Peggs have, in the
same work, given us the following instructive information :—

'Deep wells are much less liable to contamination
than shallow wells, but even they are not safe from the
insidious influence of cesspools. A very striking instance of this occurred at Liverpool some years ago in
the case of the Dudlow Lane well, sunk in the New Red
Sandstone formation. This well was situated in a suburban district some distance from Liverpool, and was
247 feet deep with a bore hole at the bottom, another
196 feet deep, making 443 feet altogether. The effect of
the continuous pumping from this well was to dry the
wells of the houses in the neighbourhood, and these were
then used in several cases by the householders as cesspools. The consequence was that the water in Dudlow
Lane well was gradually polluted, and in five years after
the well was constructed it had to be disused.'

The following is the official report of the Water Committee on the matter :—

'In the case of the Dudlow Lane well the committee
were compelled to cease pumping from February 1872 to
May 5, 1873, in consequence of the dangerous extent to
which the water was contaminated. It was ascertained
that the evil was mainly due to percolation from cess-

pools and disused wells which had been receptacles for drainage; and the committee caused the communication with several of these to be temporarily diverted, at the same time pressing the local authorities, and co-operating with them, to carry out a complete sewerage scheme for the district. By these means the quality of the water was so far improved that it was brought within the limits defined by the Rivers Pollution Commission as "reasonably safe," and the pumping was resumed.'

Now we are told that the danger of contamination of deep wells is not due merely to their depth but still more to the depression of the level of the water which is caused by the pumping. When the demand for water is great and the pumping is severe the flow of water from the soil around the well into the well itself is considerable, as the distance to which the influence of the pumping extends depends so much on the depression of the water due to the pumping; it is convenient to express this distance in terms of the depression, or, in other words, to say that the distance the well draws is twenty times the depression, thirty times the depression, and so on. This distance is most important from a sanitary

Locality	Authority	Nature of strata	Depression of water in well	Extreme distance to which influence of pumping extends	Ratio of distance to depression
			ft. in.	ft.	
Nuremberg(1)	Thiem	Fine sand	1 4	33	24
,, (2)	,,	,,	2 2	33	15
Dresden	Salbach	Fine gravel	8 2	108	22
Leipsic	Thiem	Very coarse gravel	6 7	1,050	160
Gravesend	Clutterbuck	Chalk	10 6	600	57
Liverpool	Deacon	New Red Sandstone	82 0	11,710	143

point of view, determining, as it does, not only whether one well will influence another, but whether or not a well will be polluted by a cesspool or other source of contamination in the neighbourhood, and we have therefore given a few examples of the distance under different circumstances :—

From the foregoing table it will be seen 'that the distance to which the influence of pumping extends varies greatly in different cases, being in one case only fifteen times the depression, and in another as much as 160 times the depression. The chief circumstance which seems to influence the distance is the degree of permeability of the strata through which the water has to percolate. In fine sand and fine gravel, where there is a large amount of resistance to the passage of the water, the distance varies from fifteen to thirty-nine times the depression. In the chalk, where fissures exist which facilitate the passage of water, the distance is fifty-seven times the depression.'

Thus it appears that even the deepest wells may be fouled by cesspools, and if by cesspools, equally by leaking drains or sewers. And since we cannot know when a cesspool or a deeply buried sewer begins to leak, it is impossible to feel quite secure with regard to water-supply from deep wells in the chalk, which is just now the most popular source for water, and is being largely recommended. The danger of contamination from a distance is (in the case of chalk and other porous soils) proportional to the depth of the well, and also in some degree to the demand made upon the well for water; so that a public well may not prove dangerous until population has increased around it and the demands made upon the well have proportionately increased.

With regard to the pollution of wells, it must be

pointed out (1) that such pollution is almost always from *underground* sewers or cesspools, the contents of which have had no chance of aeration and purification; (2) that such pollution is caused by *leakage—i.e.* the direct irruption of water from cesspools, &c., into wells without filtration of any kind; and (3) that water in sewers and cesspools, when it begins to leak, is under the pressure of a superincumbent column of water, and hence the burrows which it gradually forms extend long distances. The mechanical conditions of water thrown on the surface of the ground and water stored in underground tanks are entirely different—as different as are the biological conditions.

Let us look at the possibilities of a typhoid epidemic from the pollution of a public well. At Lausen we saw that 19 per cent. of those who drank the polluted water suffered from typhoid, and if the same proportion be maintained in other instances, then we might reasonably expect, in the case of typhoid poison finding its way into a public well, that nineteen would suffer out of every 100 persons dependent upon that well for their water—190 persons in a population of 1,000; 1,900 persons in a population of 10,000; or 19,000 persons in a population of 100,000. Of those attacked 5 per cent. at least would die.

Water companies throughout the country ought to be made liable to an action for damages in the case of their water being turned to poison. The most vigilant supervision must be maintained in order to prevent contamination; and frequent analyses should be made by independent analysts at the expense of the company.

It is, we have seen, an easy matter to foul rivers, watercourses and wells. How can water be purified when once fouled? *Filtering* only removes coarse floating impurities, and most certainly is not to be relied upon for the removal of typhoid poison. When a filter has been too long in use it may dirty the water instead of cleaning it. Filters of sand which are used intermittingly, and which are thoroughly aerated, seem to be the best, and to be able (by means of organisms growing in them) to oxidise much organic matter.

Boiling will probably destroy typhoid and similar poisons; but boiling for a short time only is, as is well known, not absolutely reliable.

Evaporation and recondensation is a sure method of purification. This is being done for us constantly by the sun, which evaporates the water which falls upon the earth, raises it in clouds, and gives it back to us again as rain. Nature is constantly engaged in purifying the water.

If we wish to have a constant supply of pure water near at hand we must religiously abstain from careless water pollution. Prevention is better than cure, and it is far easier to stop water pollution than to remedy it.

In rural and semi-rural places water should never be used for carrying excrement; and building should be so controlled that water-carried excrement may not become necessary. Water fouled by domestic use should be thrown on the nearest available piece of ground. Some will be thus evaporated, some will be absorbed together with much organic matter by the roots of growing plants; and the rest will filter slowly through the earth and find its way to a watercourse in a state of practical purity.

During the summer months, while vegetation is vigorous and the temperature high, scarcely any of the water will soak far away, but all will be evaporated and absorbed by the roots of the plants. Trees and vegetables, be it observed, are the best and really the only effectual scavengers. They also suck the water from the soil and keep it dry. It has been estimated by Pettenkofer that an oak tree with 711,592 leaves will, during the summer, evaporate $8\frac{1}{3}$ times the amount of rain falling on the ground which it covers. The *Eucalyptus globulus* will evaporate eleven times the rainfall. We should always expose waste water to the air and to alternations of temperature; the heat or the east wind will dry it up and stop the growth of organisms; cold will freeze it, and equally stop the growth of organisms. In sewers and cesspools there is neither heat nor cold, summer nor winter. In that muggy damp atmosphere evaporation and oxygenation are impossible. And waste water, after travelling miles of pipes, is not appreciably diminished in volume, and is charged in addition with whatever of impurity it may have met with in its dark journey.

From what has been said it will be gathered that cesspools ought not, *under any circumstances*, to be permitted. If they be mere holes dug in a porous soil, their contents may soak nobody knows where; if they be impermeable, they are still hot-beds of filth-disease, which affect those in the neighbourhood.

There is all the difference imaginable between a cesspool and an old-fashioned privy. The latter was more or less open, but little liquid found its way into it, and evaporation rendered the contents so solid that soakage and leakage were, if not impossible, at least difficult of occurrence.

A cesspool receives water, and its contents *must* soak away, diffusing poison through the earth. The constant pouring of liquid slops into the same hole day by day is sure to cause cracks and fissures in the soil, and the pressure of water is sure to force an outlet often where least suspected.

Cesspools must be written down as the most immoral

of all insanitary subterfuges, and their construction should be absolutely disallowed. Excrement should never be allowed to come into contact with water. Open channels are better than closed pipes for the escape of waste water from houses.

Closed sewers should only be resorted to in cases of the direst necessity and with a full sense of their danger. And solid excrements (which are often dangerous poisons) should be kept out of them lest the diffusions of excrement poison become co-extensive with the sewer.

Under existing conditions surface wells are not safe sources for water. A well of moderate depth, protected from surface drainage and in the middle of a well-cultivated plot of ground, would be a safe source for water if no cesspools existed. Surface wells in towns, the soil of which is excrement sodden, are little better than cesspools, and they are highly dangerous. The most dangerous surface wells of all are probably those in big towns like London, where, owing to the gas in the earth and the sulphur in the air, vegetation is at its minimum.

As we have seen, no method of purifying fouled water short of evaporation does anything but remove the coarse impurities. Neither domestic filtration nor filtration by public bodies can be regarded as certainly reliable for ridding water of organic poisons. The schemes which are so general throughout the country for precipitating and filtering sewage water succeed only in making the water less objectionable to the senses. They often add to the amount of *dissolved matter* in the water, and certainly leave the organic poisons untouched. All this is recognised and stated by the Rivers Pollution Commissioners in their sixth report made to Parliament in 1874, and yet we find the sanitary authorities of this country countenancing and even encouraging such schemes (notably in the Thames Valley), well knowing that, after the expenditure of millions of capital and a large annual outlay, the Thames water will be even less fit for drinking purposes than it was before. While these schemes are countenanced, be it observed, there is no attempt to make individuals do their duty.

The idea seems general that it is impossible to supply too much water for the daily use of households. This is very questionable. Enough is as good as a feast. No very large amount of water is needed for the attainment of absolute cleanliness, both personal and domestic. The man who is minded to be clean will attain his end with a small amount of water; and even though we take a river to those who love dirt, they will make no use of it.

It is certainly not advisable to dirty more water than is necessary, because by law the water must be purified again before it returns to the river, and this entails endless expense on sanitary authorities.

If those who rely on public bodies for their water-supply are made to pay for exactly as much as they use, we may be sure that no excessive waste will take place, and there is but little fear that the price will be such as to prevent even the poorest from having enough.

The objection which is raised to the supplying of water by meter is, that under such circumstances the poor would be insufficiently supplied. It would be easy, however, to adopt a sliding scale of charges, giving the water of necessity at a low rate and charging more for the water of luxury. If sixpence per thousand gallons were charged for the first ten gallons per head per diem, this would amount to 1$s.$ 9$d.$ per head per annum. A shilling per thousand might be charged for the next five gallons per head per diem, and 1s. 6d. for the next five gallons, and so on.

In towns the question of water-supply assumes an urgency which is proportioned to the degree of overcrowding, and it is in this connection that it becomes advisable to say a few words on what may be spoken of as *Hygienic units*.

The prime hygienic unit is necessarily the individual man, and the problem which sanitarians have to solve is how to provide this individual with pure air, pure water, food, and raiment. The individual requires a definite average amount of pure air, a definite average amount of pure water, and a definite average area of the earth's surface for producing his food, clothing, and other necessaries. It may not be unprofitable to consider these units of air, water, and earth in relation to the individual, because it will be evident that our ability in the present day to practically neglect one of them causes serious difficulties in dealing with the other two.

Human life is only possible on the condition that a certain area of the surface of the earth be dedicated to its support.

Pope has drawn a charming picture of the recluse,

> Whose herds with milk, whose fields with bread,
> Whose flocks provide him with attire,
> Whose trees in summer yield him shade,
> In winter fire.

And since all the necessaries of life come from the earth, it seems but natural to ask the question: What area of ground is necessary for the support of a man?

WATER

Although this question must be fundamental when discussing matters of practical hygiene, it is, nevertheless, not capable of any exact answer.

If we speak of the area necessary for the support of an individual as an earth unit, it will be at once evident that the unit must vary in size with the fertility of the ground and the latitude in which it exists, because more food, firing, clothing and housing is required in cold climates than in hot ones. It would be very hazardous to make any calculations for practical application, because the fertility of the soil varies, and the unit which might be sufficient in a year of plenty would fail to support its owner in a year of scarcity.

Theoretically, an island like England should support a very large population, because, being surrounded by the sea, an inexhaustible source of food is within reach of all.

A soldier on service is said, according to Parkes, to need about 31 oz. of dry food per diem, of which 18 oz. should be carbohydrates, 7 oz. proteids, $4\frac{1}{2}$ oz. fats, and $1\frac{1}{2}$ oz. salts. Now, 30 oz. of oatmeal contain about the requisite quantity of carbohydrates together with $3\frac{1}{2}$ oz. of proteids and $1\frac{1}{2}$ oz. of fat. If, therefore, a man had 2 lb. of oats per diem, and were allowed fish *ad lib.* for supplying the deficiency of proteid and fat in the oatmeal, it is theoretically conceivable that he might continue to exist and work. If we take the yield of oats per acre at 50 bushels, weighing 40 lb. per bushel, this gives us 2,000 lb. of oats per acre; and if 2 lb. per man per diem be sufficient (with the addition of fish) to support life, it is evident that each man requiring 730 lb. of oats per annum needs rather more than one-third of an acre for his food-supply. Although it is well known that oatmeal and herrings formerly constituted the staple food of the Scotch peasantry, and still forms no inconsiderable portion of their diet, this third of an acre must be looked upon merely as a theoretical minimum, and is brought forward for the purpose of enforcing the self-evident fact that a certain area of ground is necessary for the support of each human life.

Although agriculture advances with civilisation and the productiveness of the soil is capable of being increased to a very decided extent, yet it is probable that the needs of civilised man more than keep pace with the improvement of the soil, and as all the paraphernalia of civilisation come directly or indirectly from the soil, it is certain that the higher the state of civilisation the greater is the area of soil necessary for the support of the man. Be it remembered that our complicated clothing and highly finished dwellings are, equally with our food, all productions of the soil.

Our theoretical minimum is calculated for a subsistence diet for a man in full work, and if, bearing in mind that a man must be clothed and housed as well as fed, we double this theoretical minimum, and treat men, women and children all alike, it is probable that this two-thirds of an acre would be a sufficient area of cultivable land for the bare support of each unit of our population.

Of the 37,000,000 acres of England and Wales, about 28,000,000 are cultivable, and as the population of England and Wales is about 28,000,000, it seems probable that, if our foreign supplies were stopped, we might, with the help of our inland and sea fisheries, manage for a time at least to support our population on home-grown produce. The Chinese are credited with supporting a very large population in proportion to the area of cultivable land in the country, and although we have very little certain knowledge, it is on all hands admitted that the population of China is exceedingly dense, that they export a very large quantity of home-grown products, of which the chief are tea and silk, and that their importation of foodstuffs is insignificant. The Chinese are probably the most hard-working, contented and thrifty people in the world, and they are probably ahead of all other nations in their knowledge of practical agriculture and pisciculture, and it is, of course, to their advance in these matters that the possibility of feeding so large a population on home productions is greatly due.

Although the human population in China is very large, the animal population is not so great in proportion. In England the horses, cattle, sheep, and pigs are collectively more numerous than the human population, but in China this is not the case. Almost all the labour which with us is done by horses or by steam is in China done by human beings, and the soil is made to a great extent to produce food for man without the intermediate action of live stock. These facts in a large degree explain how it is that such a numerous population is sustained. It also explains the opposition of the Chinese officials to the introduction of railways and steam machinery. Almost the only source of force in China is human muscle, and as this astute race is not likely to be deluded into the belief that steam machinery can *create* force, and as the people are contented and peaceable, and as their trade goes on steadily instead of by fits and starts, booms and strikes, they are naturally unwilling to introduce machinery which must have the effect, as it has had with us, of dislocating their industries, and must bring about a complete change in the habits of the people. The Chinese seem to adhere to the maxim, ' Let *well enough* alone.'

When we speak of the dense population of China it is

well to remember that the Chinese live upon one plane, and that the population is not piled up in houses storey upon storey as with us. This living upon one plane has the effect of bringing the population very much *en évidence*, but it is not possible to have the same amount of overcrowding under these conditions as is to be found in the leading cities of Europe and America. The city of Pekin contains about sixty persons to the acre, and when viewed from the top of the walls it is said to present the appearance of a city of gardens.

In England and America the discovery of the steam-engine has led to an excessive concentration of population in certain localities for manufacturing purposes, and in order to feed these urban populations it has been found necessary in Great Britain to remove all restrictions on the importation of food, and thus it has come about that the agriculturist has had no share in the commercial prosperity of the country; the artisan has been artificially fattened, and the agriculturist has been starved, and it is in no way surprising that the latter should wish to change from the plough-tail to the workshop, and thus increase the competition with which artisans have to contend. The land which is necessary for the support of the British individual is nowadays scattered about the globe, and may be at the Antipodes. It is usually completely out of sight, and the consumer of imported food seldom gives a thought as to how and where it was produced. Imported food is so cheap that the high cultivation of our own country has ceased to be a matter of prime importance, and there being but little demand for organic manure we have begun to burn so much of the refuse of our cities as is combustible, while the rest is used to destroy our fisheries and block our ports. What the end will be it is not difficult to see.

The fertility of a country which imports a large proportion of its food ought, if a rational use were made of excremental and other refuse, to increase, and the ability of the land to support life should increase with it, but in England there has been for years a gradual process of agricultural degradation, and corn-lands which supported human life have been largely converted into pastures for the support of the lower animals.

Owing, so to say, to the neglect of the earth unit and the extraordinary and unprecedented concentration of population in certain localities, we are beginning to experience difficulties in supplying the individual with his unit of water.

The London County Council has recently stated that any scheme for the supply of water to its district must, in order to be efficient, be calculated on a minimum supply of thirty-five gallons per head per diem for a population of 12,500,000! Not only is the Council apparently endowed with the spirit of prophecy, but it seems to contemplate a concentration of population on the area which it controls three times greater than that which obtains at present. If, instead of sixty-five persons to the acre, we are eventually to have 195, and if the Council is going to do everything to encourage and nothing to check this fearful concentration, it is evident that the public health must steadily deteriorate.

However, let us accept the figures and see what they mean. Thirty-five gallons for 12,500,000 is 437,500,000 gallons a day and close upon 160,000,000,000 gallons per annum. This amount of water is nearly one-third more than that which is calculated to flow over Teddington Weir. Having fouled one Thames until it is dangerous to drink of it, London is crying for a second. The question is, Will it get it?

This question of water-supply for our big centres of population has of late years become rather urgent, and is likely to occupy public attention for some years to come. The fact that London and Birmingham have both fixed their eyes on the same source in the Welsh hills, and that Birmingham was the first in the field and has compelled London to look elsewhere, is one to make us ponder the whole question rather seriously. Glasgow has taken Loch Katrine, Liverpool has taken Lake Vyrnwy, Manchester has cast its eyes upon Thirlmere, Birmingham is looking to the Welsh hills, and all these towns have expended, or are seeking to expend, enormous sums on the erection of municipal waterworks. This, of course, has to be done with borrowed money, and as our successors will have to pay a great part of our bills it is very necessary to be sure that posterity will reap a benefit as well as a liability.

The great aim of the sanitarian should be to prevent overcrowding, and it at once becomes a question whether, regarded from the national point of view, it is better to allow populations to settle in spots where water is to be obtained, or by the expenditure of millions on great engineering schemes to bring the water to certain spots and thereby quietly encourage overcrowding, with all the physical and moral degradation which it entails. There can be no doubt that the bringing of water under pressure in large quantities to any given spot tends to increase the rateable value and is good for the landowner and the manufacturer, but whether it is of benefit to the public health is, to say the least, doubtful.

That it is the duty of sanitary authorities to stop the pollution of rivers and other sources of water there can be no doubt, but this is a duty they practically never perform, because it is a disagreeable one and brings them into conflict with individuals.

The duty of a sanitary authority to provide pure water is interpreted as an injunction to embark on a huge speculative business with thousands or millions of borrowed capital—a speculation in which posterity runs the risk and the directors incur no responsibilities. This is a proceeding very dear to the heart of the average vestryman, and is naturally popular because it is considered to be 'good for trade.' The amount of water required per head is, in these modern times, very large. The inhabitants of London consume about thirty gallons per head, which is slightly below what is considered necessary—thirty-five gallons per head being the amount which is at present accepted as sufficient. This number is, of course, an average, and includes the water used for animals as well as man, for municipal and manufacturing purposes, and also for power, such as the working of hydraulic lifts and the driving of light machinery; and it is to be noted that the use of water as a motor is on the increase. Great as is the amount of water used per head, it is likely to be greater in the future.

Now all the water which is available for our consumption has its origin in rain which, falling upon the earth, percolates through it to appear again in the form of springs, or finds its way along the lines or surface drainage to the lakes and rivers. An inch of rain is equal (approximately) to 22,000 gallons per acre (actual number 22,624), so that an annual rainfall of 30 inches gives 660,000 gallons of water per acre. If this quantity of water falls upon cultivated land, what amount of it will percolate through the upper strata of the soil and find its way to the wells?

This is necessarily a difficult question to answer, and

must depend upon the character of the soil and the nature of the growth and crops upon it. The draining of land *upwards* by means of the roots of plants and the evaporation carried on by the green leaves is so great that no water can possibly be added to the subterranean stores during the period of growth—*i.e.* from the middle of March to the middle of October; and this fact, which is generally recognised, finds expression in the saying (common in many parts of the country), that 'the springs never rise until after the first frosts of autumn'—*i.e.* until the green leaves have been killed and their large evaporating surface abolished.

Messrs. Lawes and Gilbert, by their experiments on rainfall and percolation, carried on at Rothamsted, have shown that of the rain falling upon the bare earth the amount lost by evaporation is tolerably constant. At Rothamsted this amounts to about 17 inches per annum, so that with an annual rainfall of 30 inches about 13 inches will percolate the soil and 17 inches will be evaporated. These results, be it remembered, are those obtained with bare soil and with the soil of Rothamsted, but the relative amounts of percolation and evaporation will necessarily vary with the character of the soil and the amount of vegetation.

Mr. C. Greaves, who carried on experiments at Lee Bridge for a number of years, showed that of 25 inches of rain falling upon pure sand, about 21 inches would percolate the soil and 4 inches would be lost by evaporation; but if the same amount of rain fell upon turfed soil the percolation would be not much more than 7 inches and the evaporation would amount to 18 inches. Mr. Greaves also showed that the yearly evaporation taking place from a surface of water amounted to over 20 inches,

To estimate the effect of vegetation on percolation is necessarily difficult. The final judgment of Sir John Lawes and Dr. Gilbert in the matter is given in remarks made by the latter at the Institution of Civil Engineers in March 1891. Dr. Gilbert said: 'It was difficult to estimate exactly what deduction should be made for vegetation. A large proportion of any area they had to consider was covered with vegetation. Sir John Lawes and himself had considered that the minimum amount would average 2 inches as in the case of downs and waste lands, where there was very little vegetation; where, with a heavy grain crop or good mangel crop, there might be an evaporation of 7 inches or more. Taking the average of a large area round London, partly covered with vegetation and partly bare, over a large number of seasons, they thought that between 3 and 4 inches should be deducted from the 14 inches of percolation, so leaving 10 or 11 inches. Supposing the average rainfall to be about 30 inches, that left about 19 or 20 inches for evaporation by the soil and by vegetation. This agreed very fairly with the results of Dr. Evans and others.'

These remarks were made à propos of a paper by Mr. Thornhill Harrison 'On the Subterranean Water in the Chalk Formation of the Upper Thames and its Relation to the Supply of London;' and Sir John Lawes, in a communicated note, drew attention to the fact that at Rothamsted there was evidence that one river had disappeared, that the River Ver was dwindling, and that the wells of the district had frequently to be lowered. The water-supply in the district was diminishing, they wanted all they could get, and certainly they could spare none for the necessities of London.

It may be interesting to add that analyses of the rain-water collected at Rothamsted have given as the most recent and trustworthy results 0·248 part of nitrogen or ammonia per million of water. The extremes observed were 5·491 and 0·043 per million, the variations being dependent on the richness of the atmosphere in ammonia and on the quantity of the rainfall, the smaller deposits containing the larger proportion of ammonia. The summer rains are generally richer in ammonia than the winter rains. Estimations of chlorine and sulphuric anhydride, made at the same time, give 1·99 and 2·41 parts per million.

If we estimate that one-third of a rainfall of 30 inches may be stored for future use we shall probably be in excess of the amount. To get, however, at the exact amount is less important than to recognise the fact that water is a limited commodity, and that so soon as population exceeds a certain density it becomes necessary to go far afield for water.

If in London each individual be supplied on an average with 35 gallons of water per diem, this amounts to something over 12,000 gallons a year.

Of the 660,000 gallons which fall on each acre of ground in London, we have reason to think that not more than 220,000 gallons are stored, so that at our present rate of consumption we could not accommodate much more than eighteen persons to the acre, supposing such persons to be dependent for their water upon private wells.

Very early in the history of every city it has been found necessary to bring water artificially from the outskirts to the centre.

London in ancient time not only had the Thames running through it from west to east, but it was intersected by a number of smaller streams. We read of

Langbourne, Sherborne, Wall Brook, the Fleet River, Old Bourne, Tybourne, Ay Brook, Westbourne, and Bayswater. The names of these various streams are practically all that remains of them; the streams themselves have disappeared long since. Let us take the case of the most important of these streams, the River Fleet, which was formerly a river of some size, by which barges of large tonnage could get as far as Holborn, at the spot where the Viaduct now is. The Fleet rose at the foot of Highgate, and flowed through St. Pancras, to enter the Thames at Blackfriars. In the first place, much of its water was diverted to supply conduits for the City, and this, combined with the drainage of private wells sunk by the ever-increasing population along its course, had the natural effect of largely diminishing the bulk of the Fleet River, which came to be known as the Fleet Ditch. A part of the water diverted from its source returned to it near its mouth in the shape of sewage and surface drainage, so that the shrunken waters of the Fleet Ditch became too foul to be tolerated by human sense. It was accordingly closed in and converted into a sewer, which now empties into the great intercepting sewer on the north side of the Thames. The waters of the Fleet River are now discharged into the Thames at Barking. The same thing has happened to all the other tributaries of the Thames in the London district; their shrunken waters are enclosed in sewers and conducted to Barking, a point many miles lower down than their natural point of entry into the Thames.

The dislocation of Thames water which has taken place, and is still going on, is prodigious. Every gallon of water pumped from the Thames or from wells sunk in the Thames Basin diminishes, slightly but surely, the volume of the river.

The 150,000,000 gallons which are distributed daily by the London water companies, although not all derived from the river itself, are in reality abstracted from the Thames or its tributaries, and tend to diminish the bulk of the river. In addition to this, the rain falling on the metropolitan area (118 square miles) finds its way for the most part into the sewers, and instead of replenishing the Thames in its metropolitan course is all conducted to Barking.

Again, one must remember the enormous increase of population in the Upper Thames Valley, between Oxford and Kingston. This population is supplied with water abstracted directly or indirectly from the Thames, and this further assists to diminish the bulk of the river, although a great part of this water is returned to the river in the form of sewage. That the bulk of the river is seriously diminished is shown by the fact that it has become necessary to build a new lock below Richmond in order to carry on the navigation. The shrinkage of the Thames has been less appreciable in London than otherwise would have been the case, because of the narrowing of the stream by the Thames Embankment. Thus we have in London the demonstration that the supply of water is strictly limited, and is simply proportionate to the area which is drained by the source of water (be it well, spring, river or lake) selected for the supply.

We have in London, also, a demonstration of the old proverb that 'one cannot eat a cake and have it,' and that if we pump the water from a river or the springs supplying such river, the bulk of such river must be diminished. The commercial prosperity of London is due to its situation on the Thames, which has afforded unrivalled facilities for inland and foreign trade, and there can be no doubt that, in spite of steam and railways, the maintenance of our silent highway is still of paramount importance. We have seen that the bulk of the river between Teddington and London is already so shrunken that we are being compelled to block the way of our inland navigation by a lock and weir below Richmond, and there can be no doubt that the *débris* and sludge of the sewage of some five million people cannot fail to block the way of our outward navigation. But the more money that is to be paid for steam-dredging so much the better for that particular trade, and the more difficult becomes the navigation of the Thames so much the better for the railways. This may be 'business,' but it is not 'thrift,' and there can be no doubt that modern business and old-fashioned thrift are like the hare and the tortoise of the fable, and that, in the long run, slow and steady will win the race. This opinion is not likely to be shared by those who are under the delusion that the steam-engine can *create* force instead of merely storing or transforming it, and that the wind and tides, and that most perfect of all machines, the human hand, are unable to compete with steam.

Wells within the area of a city are so certain to get fouled by the subterraneous tricklings from sewers and cesspools that it has become the fashion to compulsorily close them. Thus it comes about that the actual area occupied by a city is not available for water-supply, and the 16,000,000,000 (sixteen thousand million) gallons of water which would be yielded by wells upon the 118 square miles occupied by London are discarded as a beverage, because they have percolated through the London soil. Big cities render their own area unavailable for water-supply, and consequently they have to look to other areas.

What must be the area of the gathering ground set aside for the supply of 160,000,000,000 (one hundred and sixty thousand million) gallons of water per annum? This must depend upon the character of the soil and the rainfall of the district. If it be sought to obtain this amount from wells in the London district, it is evident that about ten times the present area of London would be necessary, or about 1,180 square miles of country. If this amount is to be brought from a district with a big rainfall (say 60 inches per annum, of which 40 would be available for storage), then a quarter of this amount, or 295 square miles, will be sufficient. In the latter case the water will be gathered from uplands, and the ground will have to be purchased outright and thrown out of cultivation, because high manuring will be dangerous to those who drink the water. Lands have been afforested, a church and village submerged, in order that Liverpool may drink. Birmingham, Manchester, and London are anxious to imitate this example. What would be said of an absolute monarch, who, having wilfully fouled his streams, destroyed a church and village in order to form a reservoir for his aqueduct?

It thus appears that a city reduces much land to a state of agricultural degradation.

First, the land upon which the city is built becomes agriculturally unproductive.

Secondly, the gathering ground for water-supply must not be highly cultivated if the water is collected by gravitation of surface-water to a lake.

Thirdly, land is needed for the purification of the sewage, the extent of which must depend upon the amount of water-supply. This land will produce nothing but rye-grass, and is, therefore, in a state of agricultural degradation.

WATER

The preceding paragraphs demonstrate the almost self-evident truth that, in order to live, we must each of us have a certain area of land, and figures have been quoted to show that in this country about two-thirds of an acre per individual might be sufficient. If the individual live on his land he will certainly get enough fresh air and more than enough wholesome water from a well sunk in it. All refuse of every kind would be returned to the land in order to maintain and increase its fertility, and finally he might be buried in it. The life of this imaginary person might not be luxurious, but his hygiene would be complete, the privacy of his home would not be invaded by a 'Board,' and there would be no sanitary rate nor burial rate.

The person living in a city needs more land than our imaginary hermit, because in addition to the area which he occupies, and which he needs for the supply of his food and clothing, he must have land appropriated for water-supply, purification of sewage and burial.

Whether or no these figures and calculations as to the amount of land necessary for the support of the individual be absolutely correct is a matter of not much importance; but it is certainly interesting to arrive at the paradoxical conclusion that the inhabitant of an overcrowded city really requires more land for his support than the country cottager, even assuming that their needs for food and clothing be identical.

The difficulties of supplying a proper quantity of *fresh air* in overcrowded cities does not need any lengthened discussion.

CHAPTER VIII

PRACTICAL DETAILS

THE writer may fairly claim to have had considerable experience in practical sanitation. For thirteen years he acted as honorary secretary, and subsequently vice-chairman, of the Parkes Museum of Hygiene, and he is bound to admit that it was in the course of listening to many lectures by several persons at this useful institution that he became impressed with the fact that the dangers of water-carried sewage more than counterbalanced its advantages, and that it is not the interested patentee who is to be regarded as the herald of the Sanitary Millennium.

In the course of professional work, both in hospital and private practice, scarcely a day passes that he is not confronted with 'filth disease' in one form or another requiring investigation as to causation; and, further, he has had experience of practical sanitation as a citizen in three distinct places, viz. (1) In London, where he has lived for more than twenty years; (2) in a country town, where, as a small owner of houses and cottages, the sanitary question has presented itself in aspects such as are common in semi-rural districts; and (3) in a village ten miles from London, where he has occupied for five years a 'suburban villa' of the commonest type, and has practically studied the question of the sanitation of a growing suburb.

PRACTICAL DETAILS

In order that the dwelling and its surroundings may be wholesome, it is essential that all excremental and putrescible refuse be removed *every day*. To allow such stuff to accumulate for a week before removal, as is done in some places where what is known as the 'pail system' is in vogue, is quite indefensible, and I believe that a daily removal would be found easier of accomplishment than a weekly removal. The vessels used for this purpose by municipalities are often absurdly heavy, cumbersome and expensive, and, *even when empty*, are more than one man can easily move, and such as to require a cart of special and peculiar construction for their transport.

With a daily removal the vessels should be cheaper and lighter—mere galvanised buckets with lids—and of such a size that one man can carry two of them when charged with their daily quota. In this way the initial cost and the cost of dragging the deadweight of the expensive two-man pails is saved, and the pails can be moved and shifted at least four times as quickly when two can be lifted by one man than when one requires two men to lift it; and when we take into consideration the fact that a man working alone has never to wait for his mate, it is tolerably clear that daily collection is economically equal to and sanitarily vastly superior to a weekly collection.

But the stuff being collected, what is to be done with it? To this question there is only one answer. It must be buried immediately, and the nearer the ground be to the houses which provide the refuse the greater will be the economic success of the undertaking. As soon as the material has been put beneath the surface of the ground it is safe; it can neither pollute air nor water, so that the greater the proximity of the burial-ground to the houses the better. It follows as a necessity that ground used for the burial of excremental and putrescible refuse *must* be cultivated; but in this there can be no difficulty, because such ground will produce all garden crops in the greatest perfection—fruit, flowers, vegetables. There is no reason whatever why public gardens and the trees planted by the side of streets should not receive their quota of putrescible matter. If applied with care and knowledge it can do *nothing but good*. It needs hardly to be said that *no antiseptics of any kind* must be mixed with the refuse before it is put to the ground. *All antiseptics, whether they be mineral salts or tar products, render the ground sterile*, and it is certain that no practical gardener would willingly run any risk of allowing antiseptic bodies to come into contact with the roots of his plants. It may be taken as an axiom that, where antiseptics are necessary, the hygienic arrangements are bad and incomplete, and it may further be taken as an axiom that, when antiseptics have been added to sewage matters in sufficient quantity to kill the microbes in the sewage and to arrest putrefaction, such sewage has no longer any manurial value, but, on the contrary, is a source of great danger to the agriculturist. Fortunately for the sewage farmer, the methods of 'treating' sewage with antiseptics need not always be taken seriously, but are to be regarded as mere perfunctory amusements which Bumble is obliged to play at so long as the Local Government Board is looking at him.

Town sewage is necessarily of most uncertain composition, containing as it does not merely putrescible matter which is good for the soil, but trade refuse (strong acids, alkalies and the like) and an abundance of antiseptics (which are always largely used both by public authorities and private persons wherever water-carried sewage is in vogue), which, applied to the soil, can only produce results of the most woeful kind.

Each variety of refuse needs its own peculiar and suitable treatment. To mix domestic refuse and manufacturing refuse, and to imagine that any *one* method of treatment will satisfactorily purify the mixture, is absurd. As a matter of fact, this indiscriminate mixing destroys the value of the whole of it.

It must be borne in mind that, in all our arrangements for dealing with organic refuse, our aim must be nitrification and not putrefaction. The key to success is free exposure to the air.

Although the knowledge that nitrification is caused by a microbe is a recent acquisition, the practical conditions which favour the production of nitrates have long been known. The following passage from a paper by Dr. J. M. H. Munro, in the 'Journal of the Royal Agricultural Society' for December 1891, puts the whole matter very plainly :—

' Boussingault, the pioneer of the experimental method in agricultural science, was well aware of the importance of nitrates and of the reason of it, one of his earliest essays bearing the title "On the Influence of Saltpetre on the Development of Plants." As early as 1856 he had succeeded in devising a method for estimating the nitrate present in soils, and he gives us the result of testing over thirty samples. He found the nitrate in traces only, or in very small quantities, in some forest and meadow soils and soils with growing crops; in very small quantity after very wet weather in autumn, and immediately after the growth of a crop; in larger quantities in fallow soils in a dry autumn, and in largest quantities after a long spell of dry weather during the summer. In one case he gives under two parts of nitrate per million in the soil of a hop field in September after heavy rains, six hundred parts per million in the same soil in the following July after a long spell of dry weather, and thirty-three parts again in the following October. Whether his figures are strictly accurate or not, the great fluctuation in this floating capital of the soil was evidently quite familiar to him, and subsequent observers have but confirmed the general tenor of his results. Greedily absorbed from the soil by a growing crop, easily washed out of it by the winter's rains, and *accumulating or being formed in the soil* during warm and not too dry weather, and especially in fallows —these were obviously the main determining circumstances of the fluctuations.

'It is only natural that the mode of formation of such a valuable substance should be an interesting and promising field of inquiry. To the natural process of *nitrification*, as it occurs in the nitre-producing villages of India, Europe has been, and still is, largely indebted for a supply of nitrate of potash wherewith to make gunpowder. The heaps of nitre-earth found near the sites of former habitations consist of house refuse mixed with porous soil, ashes from the fires, urine, &c. After long-continued exposure to Indian warmth, lixiviation of this nitre-earth with water furnishes a solution from which saltpetre is extracted by evaporation and crystallisation.

'In 1777, when France could not import saltpetre, the Government caused to be printed "Instructions for the Establishment of Nitre-heaps," which Boussingault makes the subject of one of his essays, and his observations and drawings make it plain that before his time the practical conditions of nitrification were well known. Heaps of soil mixed with ashes and animal refuse, arranged in layers separated by loose straw kept under cover, freely exposed to air, and watered as often as possible with urine, turned and removed once or twice if practicable, furnished in the course of some months a notable supply of nitre. If treated after the manner prescribed, we learn that about 450 tons of material would in two years furnish about $4\frac{1}{2}$ tons of crude saltpetre. The watering with urine was to be stopped some months before the final lixiviation. Though earth was regarded as a purely mechanical agent, and any earth not too compact would serve, the best was known to be *that already charged with nitrate*, such as cave earth, manured garden soil, the earth in the neighbourhood of stables and refuse-heaps, &c. The necessary potash of course came from the ashes, the oxygen from the atmosphere, and the nitrogen in Boussingault's time was known to be supplied by the urine and animal matter, and to be converted by the putrefaction of these into ammonia before undergoing oxidation and combination with the potash to form nitrate.

'For the potash of the ashes substitute the lime of the soil, and for the nitrogen of the animal matter that of the decaying vegetable matter of the soil, which is slowly given off as ammonia during the decay, and it is seen that the formation of nitrate of lime in soil proceeds on the same lines as that of saltpetre in the nitre-heap. That the formation of nitrate is encouraged by warmth, by moisture, by porosity of the soil, by tillage and other operations favouring free admission of air, and by the presence of lime or potash, was as well known fifty years ago as it is to-day. But the combination of the atmospheric oxygen with the nitrogen and hydrogen of the ammonia resulting from decaying vegetation was then, and for very many years afterwards, supposed to be as purely chemical an action as the combination of nitric acid, once formed, with lime or potash to produce nitrate of lime or nitrate of potash.'

When excremental matters are buried they must not be buried deeply. The end we wish to obtain is nitrification, not putrefaction; and to this end oxygen and microbes are both necessary. Theoretically, perhaps, the best course would be to allow them to remain upon the surface, for we know that such matters, when left on the surface of the ground, quickly cease to be offensive to our senses. Practically, the best course is to bury them immediately beneath the surface, one spit or half a spit being amply deep enough. To bury such matters deeply is to court failure and endanger the water-supply.

When 'earth-closets' are in use there are certain practical points which demand attention. In the construction of an E.C. every attention must be paid to cleanliness and decency and their easy maintenance. An E.C. should never be contained within the four walls of the main structure of a house. It should either be a distinct building or should be approached by a short passage or corridor with cross-ventilation or verandah.

If possible, it should be lighted from the top. In one which was built for the writer the roof is composed entirely of semi-opaque glass, and thus the least impurity is at once noticed. The floor must be of concrete or tiles. The receptacle beneath the seat should be of galvanised iron, and may be removed from the front or by a special opening in the back or sides. Guides must be placed upon the floor beneath the seat in order to ensure that the pail is always placed accurately in the same spot. The seat should be of mahogany or some equally hard wood polished, the walls should be matchboarded and varnished, and thorough ventilation should be provided.

The earth should be stored in a bin alongside the seat,

and should be thrown upon the dejection by means of a scoop, such as is used for flour or sugar. The various mechanical contrivances for precipitating the earth are too accurate. When they are used the earth falls always upon exactly the same spot, whereas the dejecta do not. To attempt in the E.C. to imitate the mechanical details of the w.c. seems to be a mere silliness. If a scoop be used, and if the earth be stored in a bin, the emptiness of the bin and the absence of earth are at once evident on entering the closet, and there is less likelihood of the necessary replenishments being forgotten. *The earth must on no account be artificially dried over a stove.* If this be done there is great risk of sterilising the earth and stopping that microbial interaction between the earth and the fæces, which is the main thing to be encouraged. In the same way the unnecessary and expensive use of heat in manufacturing 'Poudrette' serves to diminish the manurial value of fæcal matter submitted to this process. If the earth be sifted and stored in a shed for a few weeks before it is wanted it will be quite dry enough, and it is most important for comfort that the earth should not be so dry as to raise a powdery dust when it is thrown into the pan. A small quantity of ash may be mixed with the earth, but always in quantity short of that which raises a cloud of powdery dust. Ash, be it remembered, is sterile, and although it has a useful effect in drying the dejecta and providing bases for combining with nitric acid, it tends, if used in too great a quantity, to check rather than encourage microbial action. When patent hoppers are used for shooting on the earth, the dryer and more powdery the earth is, so much the better for the patent hopper, but so much the worse for microbial action.

An earth-closet should always be provided with a

brush (of the kind known as a 'kitchen hearth-brush') in order to sweep away any crumbs of earth which may be spilled accidentally. An earth-closet should be emptied every day, and the contents should be placed in a shallow trench and lightly covered.

If earth be scarce, the contents may be placed in a heap in an open shed, in which case the microbial action will soon reduce the whole to the condition of garden earth or humus. This humus may be used again and again, and the microbial activity, and its power of changing fæcal matter to humus, will suffer no loss whatever, but will rather tend to increase by a multiplication of the microbes. The bulk of the earth which is used in this way tends to increase, but the manurial value of a given weight does not tend to increase after a certain point has been reached.

Because closet-earth has been found by chemists to contain less nitrogen than pure Peruvian guano or nitrate of soda, it has been said that it has small manurial value, but those who say so form their opinions on theoretical grounds, and not from practical experience. The writer has now had ten years' experience, and his belief is that closet-earth is as fertile as it is possible for earth to be, and if used with common agricultural skill and knowledge is capable of producing all garden crops—fruit, flowers, vegetables—in high perfection. It is not, of course, comparable in any way to the artificial stimulants so much in vogue with agriculturists, and only an ignorant person would make any comparison between them. Chemical manures can only be used in a state of extreme dilution, and if used too strong may cause incalculable mischief. In the same way 'night-soil,' which consists of partially dried excrement and urine, and is more comparable to guano than to closet-earth, is sometimes dangerous from its strength, and requires to be used with caution. Closet-earth, when ripe, is *earth*, not manure. It is simply the richest earth possible, incapable probably of further enriching, but equally incapable of doing mischief by its strength.

The estimation of nitrogen does not tell us all that we want to know about a manure. Our ignorance concerning the causes of fertility and sterility of soils is still very great. The recent discovery that the fertility of the soil for certain leguminous plants depends not upon anything which the chemist can detect, but upon the presence of microbes (almost invisible to the highest powers of the microscope, absolutely imponderable in the most delicate balance), which grow as parasites on the roots of the plants, and play an indispensable part in their nourishment, must force upon us the conclusion that, useful as chemistry has been to the agriculturist, it cannot tell us all that we require to know concerning the causes of fertility or sterility.

Speaking with ten years' experience, I have no hesitation in stating my belief that for horticultural purposes closet-earth is unsurpassed. The garden where my experiments have been made is in a low-lying situation near a stream, and the only fault which I have to find with it is that the crops have a tendency to be gross and too luxuriant. The cause of this fertility is not to be attributed solely to the nitrates in the closet-earth, but it is quite conceivable that some of the microbes, which we know swarm in the intestines of man, may play an important part in the change which excremental matters undergo in the soil, and in the interchange between the soil and the rootlets.

That organic manures ('farmyard manure') are superior to all artificial manures is now generally ac-

knowledged, but farmers are apt to complain that they
'foul the land'—*i.e.* that the undigested seeds of weeds,
grasses and forage plants which the animals have consumed are liable to germinate and choke to a greater or
less extent the crop which is being cultivated.

Man, however, is a 'cooking animal,' and therefore very few vegetable seeds are liable to pass through
his intestines and still remain capable of germinating.
Closet-earth produces very few weeds, although there appear upon ground where it is used seedlings of gooseberries, currants, strawberries and raspberries, and of
any other fruit which is habitually consumed without
being previously cooked. If these seedlings be preserved,
some of them will be found exceedingly prolific.

The best kind of earth to use with an E.C. is undoubtedly garden mould or humus, and the richer the
earth is and the fuller it is of microbes the more quickly
will all be changed to earth. If earth taken from a
considerable depth be used it is sure to be comparatively
sterile, and the change is slow ; or again, if ashes, which,
of course, are absolutely sterile, be used in large quantities,
or if the earth have been accidentally sterilised by over-
heating, the action may be very much delayed. Again,
the temperature and humidity of the air are all important
in their relation to the interchanges between the earth
and the organic refuse. It is quickest in ordinary
summer weather, when sunshine alternates with showers,
and it is checked to a greater or less extent by cold and
also by drought, for without a certain degree of moisture the nitrifying microbes fail to grow. I think one
would be justified in saying that in ordinary summer
weather the humification of excreta is accomplished in
about three weeks, but that in unfavourable weather (cold
or drought) it may be delayed to a degree proportionate

to the length of the unfavourable conditions. When what I have called 'humification' is complete, all has been turned to humus with an earthy smell and without any offensive qualities whatever. The humus necessary for the working of the E.C. has been *increased in quantity* and improved in quality, and if this be used a second time the humification of the excreta will (*cæteris paribus*) be quicker than before. This is a most important matter, but, as we have seen, it is easily explicable now we know that the change is biological and not merely chemical. It is difficult to see that anything except sterilisation could destroy this power of the earth; but it is, of course, conceivable that some undesirable microbe may get a footing, just as happens occasionally with fermentation, and upset or interfere with the normal process.

In alcoholic fermentation the process is accompanied by a very large growth of the ferment, so that the product of yeast from one fermentation is sufficient to induce satisfactory fermentation in (perhaps) 100 times the bulk of the fermentable liquid which yielded it. With ordinary care the brewer not only has no difficulty in perpetuating his fermentations, but has a large surplus of yeast to dispose of. In the same way the process of humification will go on *ad infinitum* if very ordinary care be taken to see that all the conditions are favourable to the process. The power of the earth to humify excreta is not diminished but *is increased by repeated use*, ' as if increase of appetite did grow by what it fed on.' The importance of this fact cannot be over-estimated, because with care an E.C. can be satisfactorily managed wherever a few cubic yards of earth and a dry shed to keep it in are obtainable. If the earth which has been used be stored in an open shed, freely exposed to the air and turned over occasionally, it will be ripe again and ready for use in a time varying, in this climate, from three weeks to three months, as the case may be. Mr. F. Bennet, of Marlborough, who is well known as a geologist, has conclusively shown that humification can be perpetuated by the intelligent use of a very few cubic feet of earth. This fact is of the greatest importance as showing the possibilities of the E.C. system in places where earth is not readily obtainable.

If an E.C. is to be entirely satisfactory the following points must be attended to:—

1. It must be well constructed and ventilated and must be emptied daily.

2. The earth used must be pure garden humus taken from the top layer, and not the under layer, of the soil.

3. The closet-earth must be stored in a dry shed freely exposed to the air and turned over occasionally.

The causes of failure of earth-closets appear to be the following:—

1. The employment of earth or other material which is no longer 'living.' Earth which has been overheated, or taken from a great depth below the surface, has no power of humification, and the same may be said of ashes, which are useful enough to mix with earth in order to increase its absorbent properties, but which are sterile and powerless to effect any change in the organic refuse. Real 'living earth' is unequalled as a deodoriser.

2. The admixture of liquids in too great quantity. The nitrifying and humifying microbes require a certain amount of moisture for their growth, and an average amount of urine mixed with the dejecta is no disadvantage. If, however, the closet be largely used for micturition only (as is very apt to be the case in female communities), the fluid becomes excessive, putrefaction sets in, and the closet becomes offensive. If the closet

be emptied every day, as it should be, no serious inconvenience will arise from comparative failure, and if the contents be superficially *buried* every day, as I have advised, then it matters little how much mismanagement, ignorance, and carelessness is lavished upon this safe and simple contrivance.

In the North of Europe (Denmark, Sweden and Norway) a closet is in use in which by a very simple contrivance the urine is separated from the solid matter. These closets are or were recently in use even in the city of Copenhagen, and were wonderfully successful.

In female communities it may be advisable to provide a distinct arrangement for micturition only and to allow the urine to run with the house-slops. It need hardly be said that household-slops must on no account be put into the pail of an earth-closet.

In every house there is much organic refuse provided by the kitchen, and in towns where there are no gardens or back-yards this is carried away by the dustmen and hucksters or is burnt upon the kitchen fire.

In country places the putrescible refuse of the kitchen is consigned to the hog-bucket, and that useful animal the pig converts it into pork and manure, or the poultry get the benefit of it, and thus indirectly benefit us. Supposing that neither pigs nor poultry be kept, the safe and economic bestowal of such refuse is perfectly easy. Take a piece of galvanised wire netting three or four feet wide, and with it enclose a circular space about three or four feet in diameter, the netting being fastened and supported by two or three iron or wooden stakes driven into the ground. Into this little wire enclosure throw all refuse from the house and garden which is capable of rotting, the parings and waste of vegetables and other food, the mowings and sweepings of the lawns and paths, weeds, fallen leaves, &c. &c. Such a heap as this, exposed on all sides to the air, is not offensive, and the component parts of it undergo humification. When the wire enclosure will hold no more a little earth must be thrown upon the top, and the heap must be left for several weeks freely exposed to the weather. It will settle down and diminish in bulk, and finally is entirely converted into fine garden mould suitable for potting or for enriching the soil. The final act in the management of this refuse-heap is to sift it and consign the residue to the garden bonfire. When one netting inclosure is filled a second must be formed, so that in connection with a house there must always be two heaps, one forming and the other ripening. Such heaps, if freely exposed on all sides, are not in the smallest degree offensive, and if by any accident they should become so, as by the addition of an excess of cabbage leaves in 'muggy' weather, it is only necessary to put a little earth on the top, when the odour will be at once arrested.

In a kitchen garden the cabbage leaves left upon the ground are apt to be very offensive in the autumn and winter, and the smell from this cause in the market gardens round London is at times sufficient to constitute a veritable nuisance. This nuisance ought not to be suffered, because it can be instantly checked by raking the dead leaves into a heap, and throwing a little earth over them.

All ammoniacal smells should warn the farmer and gardener that he is losing valuable material.

In a house and garden there is a never-ceasing accumulation of refuse and rubbish, and it must be borne in mind that all of it has a definite value. In dealing with such refuse the rule must be (1) that whatever is capable of rotting must be put in a heap to 'humify;' (2) whatever is not capable of rotting, but is combustible, must be burnt.

In every garden it is necessary to reserve a space all the year through upon which to burn combustible rubbish, and it is needless to say that the resulting ash is of great value as a mineral manure, and contains large quantities of potash and other bases (lime, silica, &c.). When rubbish is burnt, the nitrogenous matters (which constitute the staple food of all plants) are lost, but when it is allowed to slowly undergo humification the nitrogen is retained in the form of soluble nitrates and nitrites. It is thus evident that it is wasteful to burn organic matter which is capable of undergoing humification or nitrification. This dogma applies doubly to the cremation of the dead, in which process valuable fuel is used in getting rid of that which easily humifies, but is with difficulty burnt.

Coal-ashes are no less valuable for the soil than wood-ashes, provided, of course, the combustion of the coal has been perfect. The public use the words 'ashes' and 'cinders' as though they meant the same thing, so that it may be well to state that cinders are worse than useless as a manure, and that all ashes added to the ground ought to be either white or red, according to the nature of the coal, and that they ought to be sifted through a fine wire sieve, and whatever is combustible be used for fuel. If the cinders of a house be thoroughly sifted, it will be found that the ash resulting from a ton of coals occupies a very small space, and that the ashes of an average middle-class villa will scarcely fill a couple of bushel-measures in a year.

In every house there is a certain amount of incombustible refuse, in the shape of broken glass, crockery and the like, and this will be found invaluable whenever good bottom drainage is required, and as a foundation for garden paths.

The secret of good garden paths is to have them well sloped and drained, so that water runs off them readily, and to make them of sterile materials, so that weeds will not readily grow. If beneath the top dressing of gravel a layer of ash (not cinders) be placed it will be found that in course of time the path will grow exceedingly hard and firm. Garden paths should be *swept every day*. If leaves are allowed to lie upon paths, and if the earthworms drag them into the path, the path will soon cease to be sterile and will grow abundant weeds. A garden path should be slightly convex, well drained, made of clean and sterile materials, of which the biggest pieces should pass through a sieve with an inch mesh, and should be swept every day.

There is scarcely any form of domestic refuse which is not serviceable in the garden, and as the occupant of a suburban villa I may state that in the course of between five and six years the 'sanitary' cart belonging to the local Board has never had occasion to call. I feel that it 'robs me of that which not enricheth it, and makes me poor indeed.' I see it at the neighbouring houses taking away a variety of things having a very definite value, and I feel that the times are out of joint when a body of persons co-operate in order to keep a burglar with a horse and cart whose duty it is to rob them severally and by turns. I feel it rather a strain upon my conscience to pay rates in support of an institution which is not only useless but which literally preaches thriftlessness to the unthinking poor.

Amongst the refuse of a house must be reckoned the bones of animals used for food. These are among the things which are believed to be the 'perquisites' of certain persons, and which for one reason or another do not trouble us. Between refuse which is marketable and refuse which is 'taken away' there is a broad distinction, and everyone must be left to deal with the former according to his own particular ideas. Bones which have been used for cooking and have had all the gelatin (that 'cultivating medium' for microbes which permeates the uncooked bone) boiled out of them are singularly indestructible, and if buried in the garden remain for years to hinder tillage. They should always be burnt and should be used for manure as 'bone ash,' which is invaluable. The shells of oysters and other 'shellfish' should also be burnt, as their destruction in the soil is a very slow process.

There is one form of litter or refuse which is largely the result of 'free trade,' and which, perhaps, is more difficult to deal with satisfactorily than any other form of litter, viz. the innumerable tins and canisters in which imported food is brought to our markets. They are incombustible and indestructible, and of very small value as 'scrap' for melting down. It need hardly be said that a very poor or thrifty person will put an old tin to a variety of uses. They can be made, for instance, to answer all the purposes of a flower-pot, and I can call to mind a first-rate Swiss hotel where the excellent floral decorations sprung from a basis of concealed meat tins. Finally, the best thing to do with them is to beat them into a manageable bulk with a hammer and bury them at such a depth that they cannot interfere with tillage. They are said to oxidise and disappear very quickly when buried in this way. They form a good foundation for paths and roads when they have been subjected to a thorough pounding with a big hammer.

In concluding the present chapter I am very pleased

P

to be able to quote the opinions of a well-known engineer, Mr. Charles Richardson, C.E., of Clifton. No man knows better the aims and objects of engineering and also its limits and the inutility of attempting to disobey the laws of nature. Mr. Richardson's views of the water-closet seem to be identical with my own, and he brings his own experiences by way of illustration. The following quotations are taken from a paper on ' Sewage ' read before the Clifton Scientific Club, March 5, 1892 :—

' What is the meaning of all our fears for the sanitary state of our houses, the appointment of sanitary inspectors, the anxiety of householders about the pipes and sewers and the fixing of sewer ventilation-pipes, &c., but a full acknowledgment that there is danger all around us and in every household more or less, since the introduction of the water-closets ?

' I may now relate a striking instance of the effect of the introduction of the water-closet system into an old and healthy town.

' I was formerly the owner of an estate which lay about a mile from, but in the same parish as, an old primitive country town of about 2,000 inhabitants. My tenant was the best farmer in the district, and also a churchwarden. The living was a rich one. When the old man died a new vicar came there, who was not satisfied with the old-fashioned vicarage; so, among other improvements, he put into it two water-closets.

' Now up to this time there had been no such thing in *the town*. All the houses had old garden privies, and most of them had a little well of good water. There were no underground drains, but little open ditches carried the rain-water into the town ditch, which was also an open ditch at the bottom of a meadow. Now the town had always been very healthy, and fever was almost unknown there. When the vicar had put in his water-closets, two or three of his neighbours followed his example, and they all ran their drains into the town ditch.

'This, of course, changed its contents into *sewage*, which shortly began to give forth its characteristic odour.

'The vicar was the first to notice it, and at the next vestry meeting he told them the ditch must be covered. The smell, of course, kept on growing worse, and the vicar more importunate; but the vestrymen only stared at him, till at last he told them they *must* do it, or he would write to the Board of Health in London, and have an inspector sent down, who would *order* it to be done: it could be done for 600*l*. a mason told him. An old farmer then got up and said: "I have known this place, man and boy, for seventy years, and I have never heard a complaint till you came among us. The place is very healthy, and *I* want to know *why* we should *now* be called upon to put our hands into our pockets to such a tune as that."

'The vestrymen would not move; neither did the vicar give in. He wrote to the Board; they sent down an inspector; the ditch was condemned, and ordered to be covered, as the present law directs.

'My tenant then wrote to *me*, telling me of the facts and of the hardship to him of having to pay for this sewer. I immediately wrote to him a reply, which he might show to his vicar. I pointed out forcibly the evils he was bringing upon the town, namely, that he would bring terrible fevers there, till then almost unknown; that he would destroy all the drinking water in the wells; that his sewer would cost at least a thousand pounds, and yet was only the first instalment of a much larger outlay involved in his new system; that, for his own sake, as well as that of his parishioners, I strongly recommended him to take out his water-closets and to go back to the garden privy, but to make it a *dry* privy; that this would cost him very little, and that he would thereby save his poorer neighbours the great cost of the sewer and its further developments, as well as all the other evils I had indicated. My tenant took my letter and laid it before the vicar, who read it through twice, slowly and attentively; and then, pushing it back across the table, said: "Mr. Richardson may know better than other people; but I prefer to take my own course." My tenant then stood up, picked up the letter, and, bringing his fist down heavily upon the table, said: "I will not act one minute longer as your churchwarden, and I won't give a penny more to your church or schools, or anything you propose; for I feel certain that all those evils mentioned in this letter will surely take place;" and he strode out of the room.

'The culvert was built—it cost 1,200*l*.—and all was quiet for a year or more; and then, in writing to my tenant, I asked him how things were going on. His reply startled me. He said: "The place is full of fever, and it is worse in the vicarage than anywhere else. The parson has already lost *three* of his children, and is himself on his back and hardly expected to get up again."

'The vicar was very bad for a long time, and vowed that if he ever got up again, and placed his foot outside of the parish, he would never place it again within the dreadful place. He did eventually get well, and exchanged livings with another incumbent, accepting one of very little more than half the income.

'The epidemic of typhoid fever in the place was so bad that the London Board again sent down their inspector to report upon it. His report was that all the wells were contaminated, that more sewers must be built, and that *waterworks* must be constructed to convey *good* water to the town from the hills six miles away.

'I sold my estate shortly after this, and do not know what has been done there since 1878. But does it not appear strange that the *Law* should give its support to the mere whim of an individual, and thus enable him to bring such trouble, loss and death into a place hitherto healthy?

'The *effect* of the introduction of the water-closet into a new locality is very striking in this instance. Here we have an old market town, of over 2,000 inhabitants at the beginning of this century; a place of little change, with its weekly markets and its quarterly fairs, with its old-fashioned cottages, each with its bit of garden, and most of them with little wells of *good* water, notwithstanding the old-fashioned cess-pit privy in the garden. This had gone on for years upon years, through many generations, and yet fever was quite a rare thing in the town under these old conditions. Suddenly, the water-closet was introduced, and three years later an epidemic of typhoid fever swept the place, which was so bad and fatal (particularly in those houses which had connections with the new drains) that the central Board in London sent down a special officer to report upon it!

'I have been personally connected with other cases, all of which indicate the same result, though not perhaps in so striking a manner; and I am fully convinced that the water-closet system must be abolished before long or we shall have no wholesome water left in the moderately populous parts of England. If this is so, the sooner we set about the reform the better it will be for every one, and the sooner we shall have our beautiful water-springs and pure rivulets restored to us.

'The question may naturally be asked now, "What do you propose as a substitute for the water-closet?" My reply is: "Follow Nature." She is the only safe guide; use the wonderful surface soil she has provided for the purpose, and get *utterly* rid of sewage!

'To begin with, the *dry* privy cannot easily be excelled if properly constructed. Taking first the case of the working man having a bit of garden, for whom the arrangement must be practically self-acting. The plan of construction is this: As no cess-pit is allowed, the *catch* is made by raising the privy *floor* 16 or 18 inches, and you go up two steps to enter. The back wall is built against a garden bed, and with a low but wide archway at the bottom, leading into the catch. The seat above is fixed against this back wall. The catch is floored with tiles, or bricks on the flat, *smoothly laid in mortar*, and this floor is carried through the little archway, thus extending the catch up to the garden bed outside, towards which it has a slight fall of an inch, so that all moisture runs down to the garden bed. The floor above, of the privy itself, should slope slightly the other way and towards the door.

'The privy is made fit for use by simply tipping a barrowful of earth against the archway into the catch. If the privy is now used for twelve months there is no bad smell, for the droppings remain *dry*, and there is no sewage; but it should always be cleaned out once a year, when the garden is dug. Now, the labouring man *will* dig up his garden bed for the sake of growing his own vegetables, and he will also take a little trouble to get *manure* for it when he is digging it up, and this he can do so easily by first casting the little heap of earth upon the droppings inside and then raking or shovelling them all out together off the tiled floor on to the garden bed, which can be done in three minutes from a catch so constructed that he will do it for the sake of the advantage he gets to his crops.

'Then, for a better class of houses, the same *principle* of construction must be carried out; but, in addition, a box of dry earth may be kept in the privy and a small handful of it thrown down the seat after use; this will make the privy unfailingly sweet and innocuous, so that it may be built as a lean-to against any outside wall of the house, and be entered through a door from the inside of the house. The dry earth may otherwise be thrown down by raising a handle, as in the water-closet, and as was practised in a dry-earth arrangement which was taken up a few years ago by an unfortunate company in Bristol, and worked at a loss for some time.

'I became connected with this company because of my strong feeling in favour of the earth system, and in opposition to the water system. I threw away some money upon it; but I learnt a good deal that was useful and practical.

'This company used to sell dry earth "commodes," made something like a big, easy, highbacked bedroom chair. The seat-board was made to lift as in a water-closet, and underneath the perforated seat below was placed an earth-box, much like a galvanised iron upright coal-box, to receive the dejecta, having a little earth sprinkled on its bottom. This earth-box was put in through a front panel. On the right side of the seat was a brass handle, as in a water-closet, and a pull at this threw a small handful of earth over what was in the bottom of the earth-box below. The back was double, and held the earth. This commode could be placed in a bedroom, and was perfectly sweet.

'In using the term "dry-earth," I do not mean earth that has been artificially dried or desiccated, but only earth dug from a garden bed in an ordinarily dry state; for in getting a barrow full it has to be riddled through a quarter-inch mesh, in order to keep out the stones.

'The company undertook to supply fresh earth, and brought back the nitrogenised earth, which was thrown into a stall of an old stable; and it was found that, if the heap of nitrogenised earth were left for a fortnight, all sign of former use had disappeared, *paper* and all, and that it was then fit to be used again; also that this might be repeated six or more times. On more than one occasion I examined some that had been used six times, and neither by the appearance, the smell, nor the feel of it in handling could the least difference be noticed from that freshly riddled; but I was informed that, when used on the garden or the farm, its valuable qualities as a manure were largely increased. The only trouble I ever heard complained of with regard to these commodes was in the prejudice of the servants against carrying the earth-box up and down.

'It must be borne in mind, however, that the water-closet arrangement has been vastly improved since it was generally established; so may we expect that the earth system would be very greatly improved after it had become largely adopted.

'Lastly, another question may now be asked: "How would you put down a system that is in such general use? By Act of Parliament?" I should reply: "No, that might not be fair; for it has been recognised and legislated for by Parliament." I should merely make the objectionable system pay for all its requirements by a local "rate" upon the closets. In this there could be no hardship. For example : Last summer I spent some weeks in a most picturesque little town in North Wales, where some of the best houses were beginning to introduce the water-closet. Now in the centre of that little town there was a beautiful fountain of pure water, erected by some benevolent individual for the benefit of the townsfolk, who were constantly filling their pitchers there. I could see at once that that pretty fountain was doomed shortly. When the water becomes polluted let a rate be levied upon the water-closets in the place, sufficient to procure an equally good supply until the original spring runs pure again, which it would, no doubt, in a year or two after all the water-closets had been abolished. Also let the same rate pay for all the sewers and drains they render necessary, and for the men who have to look after them. I do not think that the vicar I have before alluded to would have been so eager to put the new water-closets into his vicarage if he had known that he and his two or three friends would have had to pay for the sewer which they made necessary. At any rate, they would soon have taken them out again when the rate did come upon them !

'And now, in conclusion, I may add the following reflection :—

' This century has been by far the most remarkable, in the intellectual history of the world, for its great progress in scientific discovery and invention. But, in the midst of all the beneficial inventions made during the period, there is one which is wholly *evil*—I mean the water-closet. It bears a remarkable likeness to the eating of the forbidden fruit by our first parents. It was, outwardly, exceedingly fair to look upon, and pleasing to the eye ; but inwardly it has spread death around, and has become a veritable tree of the knowledge of good and evil! For the great spread of the diseases which it fosters has given rise to the investigations into the *causes* of those diseases and into their means of spreading. It has led to the discovery of the different micro-organisms and to the use of antiseptics.'

CHAPTER IX

PERSONAL EXPERIENCES IN A COUNTRY TOWN

In the winter of 1881-2 the writer purchased a house and about two acres of garden in the town of Andover. The house had been for some time untenanted, and the garden was so overgrown with weeds that the paths were scarcely recognisable. The house is situated in one of the most low-lying parts of the town, and is one of a type very common in country towns, being entered directly from the street, while behind it is a garden as umbrageous and secluded as can be desired, running down to the banks of a trout stream (the Anton).

Our forefathers generally chose the valley rather than the hillside as a spot for building. Valleys are fertile and sheltered, and present no difficulties in the matter of water-supply, a most important point when every gallon of water had to be raised by hand. A situation on the banks of a stream, by pure rushing water, was one to be coveted. The river was a source of power, and was always used to drive a mill; it was a source of food, for it was kept well stocked with fish; and it was a source of water which might safely be used for drinking or other purposes, and which brought fertility to the soil. In the valley life can be sustained at a lower cost than on the hill, and hence its popularity.

The nineteenth century in Great Britain has so far been an era of money-getting; the population has increased threefold, and a very large proportion have a great deal more wealth than is sufficient to purchase the necessaries of life. We have, among other luxuries of the age, the great boon of water under pressure, which enables persons to build houses in elevated positions, and which, rightly used, should in time make the hill-top blossom as luxuriantly as the valley. The right use of this water under pressure should be its application to the land round the house as it runs to waste after having served for household purposes. It would then percolate into the soil, conferring greatly increased fertility, and find its way to the valley again by natural channels, and in doing so would be a source of danger to no one. Unfortunately the water-pipe as it climbs the hill is always accompanied by its filthy companion, the sewer-pipe—that sanitary Satan which has brought 'death into the world and all our woe.' Through this pipe the waste waters return to the valley, charged with every foul abomination. This waste water or sewage periodically floods the low-lying situations, is a source of perennial offence and expense, and finally so fouls the valley-stream that none dare drink of it. Thus, under existing circumstances, those who live on high ground get a minimum amount of good from the water which is pumped up to them, while those who remain in the valley get a maximum amount of harm, and suffer financially from the depreciation of their property. This state of things must go on until it is made compulsory upon every individual householder or landowner to apply the waste waters to the land and return a clear and clean effluent to the public streams. The writer's Utopia is a place where there are water-pipes and no sewer-pipes, where every cottage on a hillside has around it an allotment sufficient to be fertilised by and to purify the waste waters, which should run clear as crystal in open channels, without needing so-called ventilation.

In country places water under pressure was a rare luxury at the beginning of the present century. Water was pumped from a well on the premises, house-slops flowed in open channels to the nearest stream, and excremental matters were deposited in a *dry* pit. The town of Andover at this time had open gutters running down either side of the street crossed by little bridges opposite the doors. With such an arrangement pollution of wells was not likely to occur, there is no record of any epidemic disease, and the river was full of fish. Every house had ample curtilage, and most of them good gardens, for the town never having been a walled fortress there had never been any necessity for overcrowding. With the introduction of the water-closet it is probable that the open gutters became unbearable, and it is certain that the old *dry* pits became converted into cesspools which endangered the purity of the wells. Between 1850 and 1860 the open gutters were replaced by underground drains. These underground sewer-drains, unlike the open gutters, have to carry semi-solids as well as liquids. Blockages are necessarily frequent in low-lying places where the drains are of small calibre and the fall insufficient. This was (and is) particularly frequent in the street where the writer's house is situated, and during heavy storms the sewage matter flowing from higher levels has occasionally been forced out of the drains and has been deposited in the street. This state of things necessarily depreciated the value of property in the street, and as it was whispered that the kitchen of the house which the writer purchased was occasionally flooded, it was not surprising that the sum ultimately accepted for the property was more than forty per cent. below the price originally named as a reserve by the vendor when it was put up to auction. Before purchasing the writer satisfied himself by a careful examination of the floors that the house was not damp, and soon after purchasing he found by ocular demonstration during a storm that the flooding of the kitchen was caused by sewer-water rushing in through *an untrapped drain.* No better example could be found of the depreciation of property in the valley by the filth flowing from higher situations. To move the kitchen sink and cut it off entirely from direct communication with the sewer was an easy matter. The house was soon let, and the writer retained about two-thirds of the garden in his own hands. This garden is situated by the River Anton, 189 feet above Ordnance datum. To the north and east the ground rises gradually until at the distance of a mile or so an elevation of 280 feet is reached. It is bounded on the west by the River Anton, on the north and east by a wall, and on the south by a double row of cottages called 'Portland Place.'

These cottages collectively form a little street leading to a private bridge over the Anton. They were owned by four or five different persons and had been built at various times, the name of 'Portland' possibly indicating that the Premiership of the Duke of Portland in 1807 was the date of the chief part of them. Many of these cottages ought never to have been built, and collectively they had two radical faults: (1) insufficient curtilage and (2) no entrance to the back premises except through the street door. In these days one hopes that no sanitary authority would allow such cottages to be built. If there be no way of removing filth except through the parlour and front door, it is certain that the evil day will be put off and that the filth will be allowed to accumulate. The collection of dirt and rubbish and the overloaded condition of some of the old privies in the backyards of these cottages are better imagined than described. These privies had been dug years previously, and the majority of them were waterlogged, because in this particular position one cannot dig a hole three feet deep without coming to water. It is essential that a privy pit should be dry. Waterlogged privies, such as these were, assert themselves in a most unpleasant manner and make a large area round them the reverse of pleasant. In some of these privies the enterprising owner had fixed water-closets, but as there was scarcely any fall to the sewer in the street, and as old boots and bits of wood and oyster-shells mysteriously found their way into the pipes, the cottages with w.c.s were perhaps worse off than those with the more primitive arrangement. The back doors of these cottages were at various levels and the cottages were divided into groups of three or four by walls. The slop-water was taken by two underground drains, one running eastward to the sewer in the street and the other running westward to a little stream the course of which is parallel to the Anton, and a few yards from its left bank. The natural slope of the surface in this situation is from the street to the stream, the latter being nearly two feet lower than the former. This underground drain was provided with the usual gratings, which were a frequent cause of petty disputes, and the drain itself was generally blocked at some point, and was a perennial source of income to the neighbouring builder. Portland Place consists of twenty-seven cottages and three other tenements at its western end, which, having more curtilage, might be spoken of as villas. These need to be mentioned, because two of them at least were provided with w.c.s which drained direct into the 'little stream' of which mention has been made.

This 'Portland Place' is close to the house which the writer purchased, and helped, it need not be said, to materially depress its value. The original object of this purchase was the garden, for amusement and the production of fruit and vegetables, and doubtless the purchaser was influenced also by a sort of sent ental regard for a spot which he had known all his life. Had this not been the case he would probably have hesitated longer before buying a property which had so many sanitary drawbacks in its surroundings.

Mention has been made of the River Anton and the 'little stream' which runs parallel to it. This latter stream rises as a spring beneath a summer-house near the centre of the writer's garden and joins the main stream of the Anton just below the first mill a few hundred yards lower down. This stream, which is only a foot or two in width, used to be a pretty little babbling brook. It received a quota of slop-water from Portland Place, but this was no annoyance whatever, as it caused only a passing turbidity. When, however, with what is called the 'advance of civilisation' two or three of the 'villas' at the west end of Portland Place adopted w.c.s, it will readily be conceived that the beauty of this little stream vanished. It became filthy and malodorous. Then commenced the covering in of the stream in the belief, which is so prevalent in the present day, that evils which are hidden are thereby remedied.

When streams of this kind are covered in it is certain that in course of time blockages will occur, and our little stream was no exception. It was covered in, in fact, and for the reasons given; it got blocked, and with the result that the garden I had purchased was improperly drained and the land more or less sour.

It was evident that the only chance of making this garden at once pleasant and productive was to get possession of the neighbouring cottages, and so to obtain complete command of the 'little stream.' Circumstances proved favourable. The property came into the market, and the writer soon became the owner of all the cottages and villas which were essential (twenty-three in all).

The various steps in the improvement of this property have been as follows :—

1. The demolition of the partition-walls between the back yards of the cottages and the provision of means of access and egress to these back premises for the purpose of scavenging.

2. The removal of nine w.c.s and the filling up of all the old privy pits and the substitution for these of an arrangement of pails which might be used on the earth system.

3. The removal of the underground slop-drain and its replacement by an *open* gutter made of Staffordshire brick running from east to west—*i.e.* taking the course of the natural slope of the ground.

4. The opening up of the 'little stream' in every part of its course, and the removal of the blockages which had occurred in two or three places.

This was followed by a permanent fall of the level of the stream in the writer's garden amounting to nine inches!

5. The providing for the daily scavenging of the cottages.

The last measure has been the most important of all. A man was engaged to act as scavenger and under-gardener whose duty it has been to remove the closet pails *every morning* and bury their contents superficially in the garden. For such a plan to succeed it is essential that the removals should be daily, and the writer has elsewhere entered fully into the economic aspects of *daily* versus *weekly* removals.

The success of this plan has been complete.

1. There is no accumulation of fæcal matter near the cottages.

2. There is no dangerous conveyance of such matter for long distances either by pipe or any other means, and the land for burial being close at hand, the daily cleansing of these cottages is effected easily and quickly, the whole process not occupying more than an hour. For the success of this plan it is essential that the cottages should be in close proximity to the land. They cannot be too close, while every yard of porterage adds to the expense.

3. All nuisance has been absolutely stopped. The cottages are kept clean, and the garden where the fæcal matters are buried is not made in the least unpleasant. The material is placed immediately below the surface—*i.e.* it is just hidden by a layer of earth. The eye sees nothing and there is absolutely no odour. In fact, there is very much less annoyance than is caused by ordinary dungings and mulchings.

4. The 'sanitary arrangements' of the cottages being of the simplest, the incessant dribble of money for the repair of 'closets,' traps, syphon-bends, gullies, ventilator-pipes, &c., has ceased.

5. The fertility and beauty of the garden have been enormously increased, and its value, which was depreciated by its filthy surroundings, has probably rather more than recovered.

It may be well to state that the garden produces all the ordinary fruits and vegetables in a state of perfection which is clearly above the average. The improvement in its fertility has been steady and gradual from the first. When first occupied by the writer it had been neglected for about two years, and, as has been stated, it was so overgrown with weeds that the paths were scarcely recognisable.

The burning of the thick felt of weeds with which the ground was covered resulted in about a thousand bushels of ashes. To reduce this wilderness to a condition of decent tillage has necessarily taken time. Nothing, probably, has been of more assistance in this direction than the removal of the blockages in the 'little stream' and the consequent lowering of the level of the water. In the first year the garden was manured entirely with stable-dung, but since the acquisition of the cottages the only stable-dung which is allowed to come into the garden is a quantity sufficient to make a hot-bed in the spring.

At the present time (and for some few years past) the garden is receiving the daily scavenging of twenty cottages with an average population of at least one hundred persons. The area actually under cultivation amounts (exclusive of paths and grass) to about 5,400 square yards, or $1\frac{1}{8}$ acre. Not only is the excremental matter from these cottages removed to the garden but the ashes as well. On making a visit to one of the cottages one day the writer encountered an ash-heap close to the back door. The heap consisted of cinders, ashes, potato peelings and similar refuse, bits of paper, fish-bones, &c. The flies were buzzing over it, and it was distinctly malodorous. A woman who had been confined the day before was lying in the next house.

'What do you do with your ashes?' was the question asked. 'Oh, we sell them to Mr. So-and-so' (the higler), was the reply. 'What does he give you for them?' 'A penny a bushel.' 'So will I, and I will provide you with a bushel measure to store them in, and when the measure is full they shall be removed and you shall be credited with a penny.' This plan has answered well. The ashes when received are carefully riddled through a sieve, and the cinders go a good way towards the maintenance of the greenhouse fire during the winter. The real ash is applied to the land, and has a most beneficial effect not only by providing mineral manure, but also by improving the physical condition of the soil. It is often stated that while 'wood ashes' are good for the land, 'coal ashes' are of small value, the reason of this distinction being caused by the confusion in the mind of the public between *cinders* and ashes. Cinders are of no use whatever as manure, and only serve to hinder tillage. There is nothing in coal *ash* which can do any harm and much that will do good, especially in close soils, to which the gritty particles of coal ash give a certain porosity.

I have said that this garden of $1\frac{1}{8}$ acre has been manured with the refuse of about one hundred persons for some years, and it may be stated that, proceeding methodically, it takes four years to go completely over the whole of the ground in cultivation. The observer is usually astonished at the small amount of excremental material which has to be dealt with, not more, usually, than will lie in a furrow ten or twelve feet in length made in the ground with a spade. Directly it is deposited in the furrow it is lightly covered, and there is an end for ever of any offence or any danger. The first crop taken off the land is always a succulent green crop of the cabbage tribe, and the plants are dibbled in on the third day after the deposit. No other crops except cabbages seem to flourish in the fresh material, but the cabbages may be followed by potatoes, these by celery (planted between the rows), the celery by peas or beans, and these again by parsnips or carrots, without any fresh manuring, and with a most abundant yield. There is no doubt that this excremental refuse confers a fertility upon the soil which is not exhausted for years. I have been urged by some practical gardeners not to apply the material to the ground at once but to store it in a heap with earth and ashes to allow it to 'ripen' before applying it. Those who give this advice have derived their experience from 'night soil' from privy pits which has undergone a certain amount of desiccation by the draining away of the fluid matter, and which is undoubtedly a most potent and dangerous manure when applied pure without previous admixture with earth and exposure to the air. By immediate burial before ammoniacal decomposition sets in there is no danger of this kind, and one is sure that nothing is lost. Further, if excrement be left above ground, blowflies and other insects will deposit eggs in it, and then the gardener will complain that 'closet-earth brings grubs,' but by immediate burial this drawback is avoided. Many practical gardeners who have seen the results of the plan of operations which has been described have admitted that the results could scarcely be better than they are.

Not only vegetables but all the ordinary garden fruits are produced in high perfection. A large contribution is always sent to the local flower show, the object being not so much to show single specimens of this or that forced to unnatural dimensions, but to demonstrate that the garden will produce every kind of common flower, fruit, and vegetable in a condition above the average, and that sanitation may be both complete and profitable.

The past season (1892) has been a favourable one for low situations because of the small rainfall and the large amount of sunshine, and the yield of the garden has been very large indeed. The gooseberries, raspberries, and currants were remarkable alike for size and quantity; one cluster of white currants (*i.e.* a collection of bunches growing from one spot on a stem) was found to weigh 14 oz. Thirteen varieties of potatoes (twelve of each sort) were exhibited at the local show, and the 156 tubers averaged more than $\frac{3}{4}$ lb. each, and the total yield of potatoes was at the rate of 11 tons to the acre. Peas, beans, carrots, turnips, parsnips, lettuce, cabbage, celery, artichokes, apples, pears, plums, peaches, are all something more than creditable for size and total produce. As a flower garden it is no less successful. For roses it is especially favourable, and the broad green path down the centre flanked on either side by hardy perennials and common annuals shows a wealth of colour and a luxuriance of growth which are almost tropical. The figures of peas, currants, raspberries, and gooseberries, which have been produced from photographs, and are one-fifth less than natural size, will convey some idea of the luxuriance of the crops (see figs. 3, 4, 5, 6).

I have dwelt upon the quality and variety of produce from this garden of little more than an acre manured with the excremental and other refuse of one hundred persons, because in this respect it is in contrast with a 'sewage farm,' which, as is well known, can be made to produce practically nothing but rye-grass and mangel. In the garden there is no excess of fluid with the manure, and the careful hand tillage brings about the aeration of the humus, and thus the nitrification of the organic matter is quickly produced. If one had to deal with the same amount of excremental

FIG. 3.
GOOSEBERRIES AND RASPBERRIES ($\frac{1}{5}$ less than natural size).

FIG. 4.
GOOSEBERRIES ($\frac{1}{5}$ less than natural size).

FIG. 5.
BLACK AND WHITE CURRANTS ($\frac{1}{5}$ less than natural size).

FIG. 6.
PEAS (⅓ less than natural size).

matter in the form of 'sewage' it would be accompanied by a daily quota of 1,000 or 1,500 gallons of water, a large proportion of which would have been previously boiled, and thus deprived of its oxygen and other gases. Under such circumstances proper tillage and nitrification is impossible, and the attempt to produce any crop other than rye-grass is labour in vain.

A word of caution needs to be uttered as to the *great harmfulness of antiseptics* from the agricultural point of view. Speaking broadly, it may be stated that any organic matter which has been mixed with chemical antiseptics or disinfectants becomes sterile and poisonous to plant life. In 'town sewage' there is always great danger to agriculture from the admixture of antiseptics, and the danger is not always to be avoided even when the 'pail system' is employed.

A town surveyor was recently contending with the writer that fresh excrement was not a good manure, and in support of this he instanced a failure on his part to grow mangel. The first year, he said, scarcely a seed sprouted, and even in the second year the crop was very stunted and miserable.

As, however, I elicited the fact that his material was mixed with a preparation of carbolic acid, the failure of his agriculture is not surprising.

If daily removal be resorted to the admixture of antiseptics is unnecessary, and it cannot be too strongly insisted upon that any such mixture of antiseptics with manurial matters is fatal to any scheme for their utilisation.

At the present time (October 1892) there are rumours of cholera, and as various schemes for disinfection have been put forward in the public press and elsewhere it may not be amiss to say a few words on the subject.

In dealing with all organic refuse we ought always to have before our eye those processes which we may call 'natural,' whereby the organic matter is nitrified and dissipated, and we should be careful to do nothing to hinder such processes. The admixture of organic refuse with antiseptics (whether salts of mercury or tar derivatives) hinders its ultimate dissolution by killing the microbes in the soil and in the organic matter itself. The admission of these antiseptics into sewers or cesspools renders the whole of the material in the cesspool or sewer a dangerous application for agricultural purposes. These antiseptic bodies, when used for disinfecting excremental matters, are often applied hastily, and there is no attempt to thoroughly mix the antiseptic with the material to be disinfected. These two are never allowed to remain long in contact, but the handle is pulled and away it goes to be diluted instantly to a degree which probably extinguishes its antiseptic power. Many of the antiseptic bodies (notably the mercurial salts) have the power of coagulating albumen, and it must be a question as to whether this action may not have occasionally a preservative effect on noxious microbes by causing spores to be hermetically sealed, as it were, in a case of coagulated albumen. If this be so, it is conceivable that antiseptics clumsily and perfunctorily used may actually preserve an infective particle. They will most certainly arrest the process of nitrification and dissolution in the earth whereby the noxious microbe is probably destroyed for ever.

The writer's opinion is that, if infective material be buried in the earth (without admixture with water or antiseptics), it is hardly conceivable that any harm can result, and all evidence tends to show that the living humus will effectively protect the wells from infection (if these be properly made) by completely arresting the passage of microbes. The fact that chemical antiseptics often coagulate albuminous fluids, and are certainly fatal to the biological processes in the soil, is one which must be ever before the mind.

The safest disinfectant is heat, and it is the only one which, practically speaking, is always at hand. On the occurrence of a case of cholera in a rural district, I think the safest course would be to bury all the excreta of the patient, and then wash the vessels and clothing in water gradually brought to the boiling-point, which water should ultimately be thrown upon the surface of the soil. The clothing and infected linen should be placed in a washing ' copper ' with cold water and soda, and allowed to soak until the albuminous stains have been dissolved. The copper fire should be then lighted and the contents thoroughly boiled. To burn good linen because a cholera patient has used it seems to be at once needless and silly. In times of infective disease, the washing copper, intelligently used, is the best antiseptic, and the cheapest, because in this way the first stage of the laundry work is accomplished, and there is no bill for poisonous chemicals. Infected clothing should not be mixed with salts of mercury or carbolic acid, because the albuminous matter (blood, &c.) is thereby coagulated, and the proper cleansing of the clothes in the laundry is interfered with. Neither should linen be plunged into *boiling* water for the same reason, but, as advised, it should be allowed to soak for some hours in cold water and soda, whereby the albuminous stains are dissolved, and then be gradually boiled. If the actual boiling of the excreta of cholera patients could be arranged before being allowed to flow into the sewer, no more effectual process of disinfection could be conceived, although, without special precautions, such process would be necessarily offensive.

We constantly hear in the present day of the dangers arising from a 'foul soil,' and there are those who attribute many of the evils which come upon us in the form of disease to foulness of the soil.

It cannot be too strenuously asserted that there is only one sure and certain way of keeping the ground sweet—viz. by tillage, aeration and cultivation of it. To pour antiseptics upon it, or to cover it with concrete, is no cure. These methods

> will but skin and film the ulcerous place,
> Whilst rank corruption, mining all within,
> Infects unseen.

If the soil be tilled so that it brings forth trees or herbage with green leaves, the air will be freshened and purified by the oxygen given off by the leaves, and the soil itself will be cleansed from all impurities. It is a common mistake to bury offensive things too deeply. If organic matter be buried in the barren subsoil instead of the living earth of the upper strata, nitrification will be delayed, and the possibility of contamination of the neighbouring wells will be greater than when such material is buried superficially, and within reach of the husbandman's tillage.

In the burial of the dead the mistake has been made not only of hindering the dissolution of the body by every means that ignorance, superstition and self-interest can suggest, but also by the fashion of burying too deeply. The body should be laid in the humus and should be covered by a mound of earth, which in all cases should be planted with shrubs and trees suitable for the soil. If this be done, it is hardly conceivable that any poisonous or other organism can find its way to the wells.

CHAPTER X

PERSONAL EXPERIENCES (continued)—WATER-SUPPLY

THE water-supply of the writer's cottages is worthy of more than passing mention.

The house, garden and cottages, as was stated in the previous Chapter, are close to and only a very few feet above the level of the River Anton, and running parallel to the Anton is a tiny rivulet which may be called the 'little stream.' Those who have not seen them can hardly imagine the beauty of the Hampshire trout streams, with the rippling of clear sparkling water over waving weed and bright patches of gravel where the trout lie. Unfortunately there is no truer proverb than that 'Familiarity breeds contempt,' and it certainly is true that the dwellers by the banks of the Anton, far from worshipping the river or even treating it with decency, seem to make every attempt to spoil its beauty and heap upon it every indignity. It is instructive on a bright summer's day to lean over the side of a boat and peer into the recesses in the bed of the Anton at points where the gardens of villas or cottages reach its banks. It is then that one realises what a pauperising thing is a running stream to those who are lazy and ignorant, how urgently our rivers stand in need of care, and how futile is the Pollution of Rivers Act!

In the bed of the Anton, one may see, by peering

to the bottom, an old kitchen range, old iron buckets and pails innumerable, hoop iron, gas-pipe, bits of wire-netting, old boots and shoes, broken bottles, crockery, old meat tins, the ribs of an umbrella, oyster-shells, and, in short, every conceivable kind of house refuse. One only sees at the bottom a very small part of what is thrown into the river, the greater part having floated away down stream. The riparian owner or occupier keeps no dust-bin and he has no rubbish-heap. He is absolutely ignorant of the right use of refuse, and throws everything into the river—dust, weeds, lawn-mowings, dead leaves, parings and trimmings of vegetables, prunings of trees, old gooseberry bushes, and finally his old garden tools and water-cans.

Under such conditions a river soon gets foul. The rubbish dams back the mud and it becomes difficult to keep the bed of the river clean. The weeds grow, and the beauty and utility of the stream are both reduced, while the thoughtless people who cause these obstructions are being deprived of material every scrap of which should be turned to profitable account. Bathing in such a river becomes unpleasant and dangerous, and boating, from the accumulation of weed, is no longer a pleasure. The trout cease to spawn when they can find no clear gravel for the purpose, and the millers complain that their head of water is seriously diminished.

'What are you going to do with that?' I said one day to a helper in the garden who was making for the river bank with a basket of weeds and rubbish in his hand.

'Chuck it in the river, sir,' was the reply.

'What would you do with it if there were no river?' was my next question, and this brought the man to his senses; and from that time I believe not a single weed or any other refuse has been thrown from my garden into the

Anton. The gardener carefully returns all rubbish to the soil either in the form of 'humus' or ash, because he recognises that low-lying gardens are much in need of replenishment, and that if everything be taken off his land and be thrown into the river the level of the soil will ultimately sink.

Is a running stream of no use to the agriculturist? River mud is most certainly a valuable addition to light soils, especially those on the chalk, but before being applied to the land it should be allowed to 'ripen' in a heap with other rubbish. Applied direct it will choke the pores of the soil and do mischief. Again, is river-weed of no use as a manure? In Cornwall sea-weed is rightly regarded as a most valuable manure, and it is certain that river-weed must have a similar though inferior value, and would do nothing but good to the light chalk soils of Hampshire.

At present a running stream is a pauperising medium to those who dwell upon its banks, instead of being of distinct value not only as a source of water and fish, but also of mud and weed suitable for putting on the land.

The recognition of the utility of a river by the agriculturist seems to be the surest road to its proper conservancy.

The above digression has been made to show the utter disregard of the average riparian for the purity of his stream. The smaller the stream the less regard is had for it, and, as has been previously stated, the little stream running parallel to the Anton was utilised to receive the contents of w.c.s and house-slops until it became a veritable nuisance, was in great part covered in, got blocked, and ceased to act as an efficient drain for the land. The history of this rivulet was, in short, that which is common to nearly all the rivulets in the country.

The surface wells for the supply of the cottages were

close to the back doors, and being only a few feet deep it was customary to dip water out of them. The dipping utensils might be anything which was handiest, and might be clean or otherwise. The parapets of the wells were of wood (in places rotten) and only a few inches high, and the paving round them was a rough pitching of the worst description, generally sodden. An underground slop-drain, often blocked (the gratings of which had that odour which is called 'faint'), ran within a few inches of the wells, and finally the steyning or brick lining of the wells was very deficient.

Under the circumstances it appeared to the writer that it was incumbent upon him to 'lay on' water from the works of the Andover Water Company, which have since (as is the fashion of the day) been bought by the municipality and paid for by a bill on posterity. This water comes from a well about 90 feet deep in the chalk, and had been pronounced of good quality. There might be fissures in this chalk well through which sewer or cesspool water might gain access to it, and the water might be contaminated by leakage from a sewer *en route* (no one could speak with certainty on this point), but the laying on of the municipal water relieved one of responsibility, and this probably was the chief reason for this course of action.

It has been mentioned how all the w.c.s were removed and how underground drains were replaced by open gutters, and there can be no doubt that the wholesomeness of the cottages has been enormously improved thereby. There is no longer any foul excremental matter in the 'little stream' (which receives only house-slops), and there are no longer gratings emitting faint or foul odours at the back doors of the cottages.

It is often stated that as house-slops have to be got rid of, and as house-slops are very foul smelling, it is the best course to construct an underground drain for their reception and allow it to carry off excremental matters as well. This is one of the chief arguments in favour of a comprehensive water-carried system. But there can be no doubt that the admission of excremental matters is the main cause of all the difficulties of the 'drainage' question, because when excremental matters are mixed with slops the mixture is so abhorrent to our senses that it *must be covered up*, and it is put underground in closed channels instead of being delivered upon the surface of the soil, in accordance with those true scientific principles which should guide us in these matters.

Now, as soon as all access of fæcal matter to the 'little stream' was stopped it became possible to open it in its entire length, to get rid of the accumulations of mud, and, as has been mentioned, to restore the proper drainage of the land under cultivation. The stream being open, any accidental arrest of the current is seen instantly and removed at once.

The stream receives the house-slops of twenty-three tenements, and I can state emphatically that this causes practically no annoyance. When a wash-tub is emptied there is a passing turbidity of the water, which quickly clears again, and there is an end of the matter. It would be perfectly easy to filter these slops through a bed of earth before allowing them to run into the stream, but they cause so little trouble or annoyance that I have not thought it necessary to do so as yet. It must be borne in mind that house-slops have been to a large extent boiled either in cooking or washing processes, and are therefore mainly sterile and far less likely to contain noxious germs than are excremental matters. Their admission to a stream is not probably fraught with any danger to the riparians lower down. This stream is kept absolutely free from accumulation by ducks, which are allowed to work in it, and are most useful in stirring the bottom and keeping it bright. Ducks are very useful as scavengers. Although the house-slops which flow into this stream are as nothing when compared with the fæcal matter which formerly flowed into it, I am nevertheless of opinion that the sanitary authority should insist upon these slops passing through some filtering medium before being admitted to a public watercourse. Personally I should rejoice to receive such an order, because it would mean that the authority had been roused to a sense of its duty as to the protection of its natural water-supplies. Public authorities are ready to spend large sums of (borrowed) money on waterworks, but seldom show the least desire to really protect the purity of streams and rivers. As a matter of fact, the town river has not been 'dragged' for fourteen years, although this is a necessary process which has to be undergone by every stream which, like the Anton, is dammed at intervals by mills.

Thus far it will be observed the measures taken for the sanitary improvements of this small property have been (1) the abolition of all w.c.s and privy pits; (2) the daily committal of all excremental refuse to the earth; (3) the replacing of underground slop-drains by open gutters; and (4) the opening up of the 'little stream' and the removal of blockages.

The weak point in all these arrangements, which seemed like a standing reproach, was this, that, although the property was on the very brink of a sparkling river which came welling out of the chalk hills, it was nevertheless advisable not to drink the local water but to obtain a supply from the water company. The shallow dip-wells close to the cottage doors were obviously so liable to contamination that, in spite of the repeated assurance of its being 'beautiful water,' common prudence made it imperative to get a supply from another source, and the 'laying on' of the municipal water was the safest and readiest plan.

The reader will scarcely need to be reminded that to 'lay on' water costs money, that the sum paid as 'water-rate' is not inconsiderable, that it has to be paid in the case of cottages by the landlord, and that it is not always easy to raise the rent in proportion. Again, cottages cannot be too simple in construction, and all sanitary fittings, including pipes, taps, &c., which are sure to be in constant need of repair, add to those expenses which the owner is bound to incur, whether or no he recover them from the tenant.

We all need to be reminded that if we, by ignorance or carelessness, foul the water which is beneath our feet, and have to bring water from a distance, the cost of living is thereby increased, rent must be higher, and less of the wages will be able to be spent in food, clothing or luxuries, and that it is a most unthrifty arrangement for dwellers in a place where the potentialities of pure water are infinite to wilfully foul this water, and go to another source a mile distant, and ninety feet below the surface. The sum paid for water for the cottages is 8 per cent. of the gross rental, and the sum paid for rates and taxes is 14 per cent. of the gross rental.

Shallow wells are universally regarded in these modern days as dangerous sources for water, and there can be no doubt that they are often contaminated, and have been a frequent cause of sickness.

How have they been contaminated? The answer to this is certain and most important. They have always been contaminated by direct inflow of filth from the surface of the ground in which they are dug, or by the leakage of cesspools or sewers direct into the well, through fissures in the soil and defects in construction.

The cause of the fouling of shallow wells is universally found in a neighbouring cesspool or sewer, in a collection of filth which has been mixed with water, and has been put into the barren subsoil, instead of being thrown upon the surface, to be dealt with by the fresh air and the living earth.

Since the adoption of subterraneous sewage methods, the fouling of surface wells has become so common that it is now the fashion to condemn them. The wells have already had, so to say, to give way to the sewer. It has been the repetition of the wolf and the lamb of Æsop's fable—the sewer fouls the well, and therefore the well is abolished. This may be practical, but it is not logical or thrifty.

If the subterraneous collections of filth were abolished, and if our surface wells were properly constructed, we might drink of them with perfect safety; for it is well known that the filtration of water through a few feet of earth deprives it of organic matter with a completeness which is almost absolute.

I determined to try the experiment of making a shallow well in the centre of my own garden—a garden, be it remembered, which is rather highly manured with human excrement.

The well is in the centre of the garden, at the intersection of two paths where I have made a circular clearance, with the well in the middle, the paths passing round it. The well is only five feet deep, and the water stands at a depth of three and a half feet. The well is completely lined with large concrete pipes, which sink into the gravelly soil at the bottom, and project rather more than a foot above the surface of the ground, so as to form a parapet. The junctions of these concrete pipes have been closed by cement, and the space outside the pipes (which are 2 feet 6 inches in diameter) between them and the soil has been filled in with solid concrete. The well has been fitted with an oaken lid, covered with lead, and tarred on the inside. In this way it has been made impossible for any water to enter this well, *except through the bottom*. The sides are absolutely impermeable to the very bottom, and the high parapet and close-fitting lid effectually prevent the entrance of rain or surface drainage. In order to draw the water a pump has been fixed, a leaden pipe from which enters the well through a hole cut in the concrete sides. This hole has been carefully closed, and the waste from the water which is pumped is taken to the 'little stream' by an iron pipe fifty or sixty feet long.

Every drop of water which enters this well must have filtered through at least five feet of earth, and, humanly speaking, it should be impossible for germs of disease to gain access to it. Within the limits of the garden of which the well is the centre, there is neither cesspool nor sewer, nor any other subterraneous collection of filthy water, and any short cut for filth from the surface by cracks or fissures in the soil has, by the construction of the well, been made impossible.

Inspection is easy, and it is possible to know the condition of the well *to the very bottom* with the greatest ease. Thorough inspection of this well is a matter of minutes, while thorough inspection of a deep well is a serious matter, and is practically impossible except to experts. With deep wells sunk in the chalk, or any other porous soil, there is always a risk of contamination by cesspools, sewers, or other similar subterraneous arrangements, the contents of which may leak into the well through fissures in the soil. One might almost say that the danger of fissures invading the sides of a well is in proportion to its depth. One may certainly say that no well sunk in a porous soil can possibly be safe, if there be sewers or cesspools in its neighbourhood, unless it have an impermeable lining reaching to the very bottom.

If filthy water were put upon the surface of the ground, all wells, deep or shallow, which were properly protected from surface drainage would be safe. If there be sewers or cesspools in the neighbourhood of a well, such well cannot be safe, be it shallow or deep. It cannot be too strongly insisted upon that the fouling of wells is due entirely to our unscientific methods of treating filth.

The only difficulty which has been experienced with the author's well is to keep insects out of it; spiders, wood-lice, earwigs, &c. finding their way beneath the crevices of the lid, and this in spite of tarring the inside. This invasion by a few insects is unimportant from a sanitary point of view, but it obviously interferes with the value of any bacteriological examination. I am unwilling to seal the well up, because thereby I should lose the power of examination and inspection.

In conversation with Professor Frankland I explained the circumstances of this well to him, and he most kindly offered to make a chemical examination. I need hardly say that I accepted this kind offer, and accordingly, in April 1892, I sent a sample of water to the Professor and in due course received his analysis and report.

The latter was as follows :—

' This water contains a large amount of saline matters in solution and is very hard.

'It is *organically very pure*' (the italics are my own) 'but it exhibits strong evidence of having been in contact with animal matter (previous animal contamination), and would on this account be condemned for dietetic use.'

This letter was accompanied by a printed circular which it is necessary to give *in extenso* in order that the reader may fully comprehend the whole question.

DRINKING-WATER

Memorandum No. 3

Previous Sewage or Animal Contamination

There is reason to believe that the excrementitious matters which exist in sewage are often possessed of intensely infectious properties; and that sewage mixing with water, even in the minutest proportion, is likely, by such properties, to spread particular diseases among populations which drink the water.

Thus is explained the peculiar power which impure waters have been shown to exercise on many occasions in promoting great epidemics of typhoid fever and cholera.

The existence of an infectious property in water cannot be proved by chemical analysis, and is only learnt, too late, from the effects which the water produces on man. But though chemistry cannot prove any existing infectious property, it can prove, if existing, certain degrees of sewage-contamination. And every sewage-contamination which chemistry can trace ought, *prima facie*, to be held to include the possibility of infectious properties.

Nearly the whole of the animal matter which gains access to drinking-water consists of sewage, that is, solid and liquid excrements.

The column headed 'Previous Sewage or Animal Contamination' in the accompanying analytical table expresses, in terms of average London sewage, the amount of animal matter with which 100,000 lb. of such water was, at some time or other, contaminated. Thus 100,000 lb. of the water of a shallow well at Andover had been polluted with an amount of animal matter equal to that contained in 5,100 lb. of average London sewage. So far as chemical analysis can show, the whole of this animal matter had been oxidised and converted into mineral and innocuous compounds at the time the analysis was made; there is, however, always a risk lest some portion (not detectable by chemical or microscopical analysis) of the noxious constituents of the original animal matters should have escaped that decomposition which has resolved the remainder into innocuous mineral compounds. But this evidence of previous contamination implies much more risk, when it occurs in water from rivers and shallow wells, than when it is met with in the water of deep wells or of deep-seated springs. In the case of river water there is great probability that the morbific matter, sometimes present in animal excreta, will be carried rapidly down the stream, escape decomposition, and produce disease in those persons who drink the water; as the organic matter of sewage undergoes decomposition very slowly when it is present in running water. In the case of shallow-well water also, the decomposition and oxidation of the organic matter are liable to be incomplete during the rapid passage of polluted surface water into shallow wells. In the case of deep-well and spring water, however, if the proportion of previous contamination do not exceed 10,000 parts in 100,000 parts of water, this risk is very inconsiderable, and may be regarded as *nil* if the direct access of water from the upper strata be rigidly excluded; because the excessive filtration to which such water has been subjected in passing downwards through so great a thickness of soil or rock, and the rapid oxidation of the organic matters contained in water when the latter percolates through a porous and aerated soil, afford a considerable guarantee that all noxious constituents have been removed.

It follows from what has been already stated that chemical

analysis cannot discover the noxious ingredient or ingredients in water polluted by infected sewage or animal excreta; and as it cannot thus distinguish between infected and non-infected sewage, the only perfectly safe course is to avoid altogether the use, for domestic purposes, of water which has been polluted with excrementitious matters.

This is the more to be desired because there is no practicable process known whereby water, once contaminated by infected sewage, can be so purified as to render its domestic use entirely free from risk.

Nevertheless, as it is very difficult in some localities to obtain water which has not been more or less polluted by excrementitious matters, it is desirable to divide such previously contaminated drinking-waters into three classes, viz. :—

1. Reasonably safe water.
2. Suspicious or doubtful water.
3. Dangerous water.

Reasonably safe water.—Water, although it exhibits previous sewage or animal contamination, may be regarded as reasonably safe when it is derived either from deep wells (say 100 feet deep) or from deep-seated springs; provided that surface-water be carefully excluded from the well or spring, and that the proportion of previous contamination do not exceed 10,000 parts in 100,000 parts of water.

Suspicious or doubtful water is, 1st, shallow-well, river or flowing water which exhibits any proportion, however small, of previous sewage or animal contamination; and, 2nd, deep-well or spring water containing from 10,000 to 20,000 parts of previous contamination in 100,000 parts of water.

Dangerous water is, 1st, shallow-well, river, or flowing water which exhibits more than 20,000 parts of previous animal contamination in 100,000; 2nd, shallow-well, river, or flowing water containing less than 20,000 parts of previous contamination in 100,000 parts, but which is known from an actual inspection of the well, river, or stream to receive sewage, either discharged into it directly or mingling with it as surface drainage; 3rd, as the risk attending the use of all previously contaminated water increases in direct proportion to the amount of such contamination, deep-well or deep-seated spring water exhibiting more than 20,000 parts of previous contamination in 100,000 must be regarded as dangerous. River or running water should only be placed in the second class provisionally, pending an inspection of the banks of the river and tributaries; which inspection will obviously transfer it either to the class of reasonably safe water if the previous contamination be derived exclusively from spring water, or to the class of dangerous water if any part of the previous contamination be traced to the direct admission of sewage or excrementitious matters.

E. FRANKLAND.

That the water was *organically very pure* was highly satisfactory, and its condemnation for dietetic purposes no more than was to be expected, and Dr. Frankland clearly could not avoid condemning such water taken, as he had been informed, from a shallow well in the centre of a garden which received large quantities of human excrement.

From the above document, however, it is clear that the well was condemned on account of its shallowness and situation, and not on the merits of the analysis. It is the almost invariable custom of analysts to condemn such water, but, nevertheless, I should have no hesitation in using the water of this particular well (1) because it is easy of complete inspection, (2) because there is neither sewer nor cesspool within reach of it, and (3) because the water is bright and palatable and is pronounced as 'organically very pure.'

For the purposes of the argument three analyses made by Professor Frankland have been placed in juxtaposition.

The first is that of water taken from a deep well belonging to the Kent Water Company.

Three Analyses of Water by Dr. Frankland, with Remarks.
In parts per 100,000.

	Total solid matter	Organic carbon	Organic nitrogen	Ammonia	N. as nitrates and nitrites	Total combined N.	Previous sewage or animal contamination	Chlorine	Total hardness
[1] Deep Well of the Kent Company. June 23, 1892	40·80	·038	·010	0	·494	·504	4,620[2]	2·4	28·4
[3] Shallow Well, Andover. Apr. 1892	37·60	·054	·008	0	·542	·550	5,100	1·9	27·8
[4] Deep Well, Andover. June 12, 1875	28·28	·106	·031	0	·444	·475	4,120	1·25	22·1

[1] 'Excellent quality for dietetic use.' 'Especially distinguished for its very high degree of organic purity.' (Report to Registrar-General, June 30, 1892.)
[2] This figure is not given by Dr. Frankland, but has been calculated by the Author from the amount of N. as nitrates and nitrites.
[3] 'Clear. Organically very pure, but it exhibits strong evidence of having been in contact with animal matter (previous animal contamination), and would on this account be condemned as unfit for dietetic use.' (Letter to the Author, May 13, 1892.)
[4] Slightly turbid, wholesome, palatable and of excellent quality for dietetic purposes. As it is derived from a deep well, the evidence of previous animal contamination which it exhibits may be safely disregarded. The hardness is rather less than that of chalk waters generally. It is well suited for the supply of a town. (Letter to Secretary of Andover Water Company, June 12, 1875.)

The second is from the shallow well in the author's garden at Andover.

The third is from the deep well formerly belonging to the water company at Andover.

The first of these waters is praised for its excellent quality and very high degree of organic purity; the second, though 'organically very pure,' is condemned as unfit for dietetic use, while the third is of 'excellent quality for dietetic purposes,' and 'well suited for the supply of a town.'

A close examination of the figures shows that the third water is probably the best, notwithstanding its 'slight turbidity' and its relatively larger amounts of organic carbon and organic nitrogen. The other two waters (the deep Kent and shallow Andover) are not at all unlike in composition, and examination of the figures only would render it difficult to say which was the best.

This term, 'previous sewage or animal contamination,' seems to me rather a misleading one (not to say terrifying). The surface of the earth is the common receptacle of dead organic matter of all kinds, and rain passing through the upper layers of the soil *must dissolve* the nitrates which are formed *in these upper layers* and carry them through to the deeper ones. For the purpose of converting organic matters into nitrates the first few inches of the soil (owing to the access of air and a plentiful supply of microbes) is of more value than all the rest, and if filthy water escape the action of the upper layers it may travel for any distance without being nitrified. The cases of the Dudlow Lane well at Liverpool (p. 168) and the typhoid epidemic at Lausen (p. 163) may be studied in this connection.

The nitrifying power of the soil increases apparently with cultivation, and there is good reason to believe that the nitrifying power of humus is proportioned to the degree of high cultivation to which it is brought. If, however, filthy water be allowed to escape the action of the upper actively nitrifying layers, and be conducted by pipes beyond the reach of them, as is the unscientific custom of to-day, it is very doubtful if proper nitrification is ever attained. The nitrates in solution show that organic matter has been nitrified and rendered harmless, and it is difficult to understand why they should be a sign of danger in shallow wells and lose such significance in deeper wells notwithstanding the fact that the deeper layers of the earth have but little action on organic matter. If dangerous filthy water has managed to leak five feet through the earth without undergoing any nitrification it would experience no difficulty in trickling in the same condition for the next 500 or 1,000 feet, and the deeper it gets the less likelihood is there of nitrification taking place.

While, undoubtedly, we want evidence of *present* sewage contamination, it is difficult to see the bearing upon dietetic value of *previous* sewage contamination. The fact that saltpetre owes its origin to the fermentations of a dung-heap in no way affects its wholesomeness.

While freely admitting that, *under conditions commonly existing*, shallow wells are dangerous sources of water-supply, one must insist that the danger is entirely due to water-carried sewage leaking from sewers or cesspools, and one must have considerable doubt whether deep wells are so uniformly safe as is generally supposed, because filthy water having escaped the action of the upper layers of the soil stands little chance of nitrification in the deeper layers. Dr. Kenwood, of the Hygienic Laboratory, University College, has kindly made two analyses of this water which do not materially differ from that of Professor Frankland. Dr. Kenwood, however, regards the water, judged on the merits of its analysis, as a 'fair water,' and reasonably safe.

To go back to the lesson which is taught by this garden at Andover. The excrement of over 100 people is buried daily in little more than an acre of ground, with horticultural results distinctly above the average. The slop-water runs in *open* gutters to an *open* stream; there is not a single underground sewer-pipe, drainage-pipe, or cesspool upon the premises. A shallow well sunk in the centre of the garden yields water which is 'organically very pure,' and which, personally, I should not fear to use for dietetic purposes,

Here we have the complete circle of changes—the right use of refuse leading to the profitable production of food without causing any danger to the water. In low-lying villages the best source of water is probably to be found in *properly constructed* surface wells, provided there be no sewers or cesspools, and slops run in *open* channels on the surface of the ground, and, of course, away from the wells.

The bacteriological examination of the writer's shallow well has been so far satisfactory. Dr. Sims Woodhead has been most kind in supplying the necessary cultivating media and apparatus from the laboratory of the Royal College of Physicians and Surgeons, and the examination has been made twice. The inoculation of the gelatine plates was made by the author on the spot. On the first occasion the well was new, and *débris* of all kinds had been dropped into it and the water had been stirred up. There were found some 700 growths per c.c., and nine species, among which mucors largely predominated. This examination scarcely merits to be recorded, because the circumstances were obviously unfair. On the second occasion, in July, there were found only eight growths per c.c., and only three species could be made out. This amounts to practical purity; for it must not be forgotten that the pump through which the water is drawn stands in the open air, and that it has not yet been found possible, as has been said, to prevent sundry small insects from finding their way round the rim of the well-cover.

This freedom from bacteria might have been safely predicted, for it has been absolutely established that humus is, as a filter for bacteria, quite unsurpassed. The clearness and brightness of this water are absolute, and when the well was last inspected (late in January

1893), every stone on the bottom was clearly visible, and no appreciable sediment of any kind had taken place.

It is not only the vital and biological condition of humus which makes it such a good filter, but its mechanical condition also. Well-tilled humus is loose, porous and crumbly, and it is not liable to have dangerous cracks or fissures formed in it as the result of drought and heavy rain. The surface of well-tilled land soon adjusts itself to the pressure of water falling upon it, and it is hardly conceivable that filth can make a 'short cut' into a well properly made and with cultivated surroundings. The best surrounding for a well is probably turf, and the next best is *well-tilled* humus.

No well, be it deep or shallow, can be always safe unless it be properly looked after and receive intelligent attention. The old-fashioned dip-well, without parapet, surrounded by coarse pitching and sour-smelling puddles, where filthy water is thrown day after day, must get fouled, and those who drink of it do not deserve to escape the penalties of their neglectful carelessness.

It is very important that the waste water from a well be conducted to a safe distance. The constant drip of water *in the same place* day after day is sure to wear a channel along the course of pipes or brickwork, and in this way a short cut by which filth may reach the well will certainly in process of time be made. It is the constant delivery of liquid filth day after day in the same place which constitutes the danger of cesspools and sewers, for directly a leakage is made it is bound to travel steadily along the path of least resistance. Those who wish to be safely guarded against leakages of this kind must be careful never to throw their slop-water for many days together in the same place. In sewers and cesspools we are apt to get a column of water capable of exercising no little pressure, which, of course, increases the danger from leakage almost infinitely. It must never be forgotten that filth in a cesspool underground and filth deposited on the surface of the ground are in totally different conditions, both vital and physical. On the surface those chemical and biological processes which constitute our protection are forwarded, and the dangerous pressure of a column of water is impossible. In a cesspool nitrification is delayed, and it is almost certain that sooner or later the pressure of the fluid will cause an irruption which must in time worm a passage to the nearest water source.

CHAPTER XI

PERSONAL EXPERIENCES IN A LONDON SUBURB

In the year 1887 the writer supplemented his professional residence with a country cottage in a suburb in the Thames Valley.

The cottage selected was a 'villa' of the commonest cockney type; abominably planned and abominably built, but not much worse than its neighbours. The attraction lay in the fact that all its windows looked south or east; that from its windows one commanded a view of an attractive rural district; and that the villa had a garden of a quarter of an acre, very picturesque and well exposed to the sun.

This little house drew its water from a private well and drained through a grating in the ground which presumably took the sewage somewhere, but no one could tell where that somewhere was, and no little difficulty was experienced in finding the inevitable cesspool. The w.c., wedged into a recess close to the kitchen door, was absolutely dark and without ventilation, except into the house. Although this house was not the author's property, the owner allowed him to carry out his ideas and so rearrange matters that all refuse should be returned to the soil without the intervention of sewer or cesspool. It must be remembered that the design of the house was, so to say, a fixed quantity, bad but unalterable, and the plans adopted were merely makeshifts, and such as proved feasible under the circumstances.

The old w.c. was routed out with its pan, levers, cistern, and pipes, and the cesspool was emptied and abolished. An E.C. was erected just beyond the limits of the house. The contents of the E.C. are buried every day just beneath the surface, and thus the question of excrement disposal was easily settled.

Next as to the bestowal of slops: (*a*) kitchen-slops, (*b*) bedroom-slops (soapy water and urine), and (*c*) the waste of a big fixed bath holding about thirty gallons.

The kitchen sink was so placed that its waste-pipe could not reach the outside of the house without being placed beneath the floor, and therefore the receptacle for this waste had to be sunk in the ground. A galvanised iron 'copper' with spherical bottom is used for this purpose, and this is emptied every day and put to the shrubs or wherever it may be wanted. This is done by the lad who cleans the boots &c., and takes four minutes at the most. It has been done by a succession of lads, and no difficulty has ever arisen. There is no smell, no unsightly appearance, and this waste water has proved very fertilising. At intervals the receptacle for the kitchen-slops is scoured and polished with sand-paper. If it had been possible to obtain a fall for the kitchen waste, it might have been feasible to empty it by turning a tap instead of by a dipper, and thus some time would have been saved; but in practice the necessity of dipping has proved to be no hindrance to efficiency, and the shape of the receiver lends itself to cleanliness. Experience leads one to say that kitchen-slops should always receive independent treatment. Of all the house-slops they are infinitely the most concentrated, and contain a large amount of grease and suspended matter, so that in arranging for their disposal the constant shifting of the spot on which they are deposited is of importance.

For the disposal of the bedroom-slops the following method was adopted: A sink was placed outside one of the upper windows. This sink has the appearance of, and is indistinguishable from, a flower-box, and is, in fact, used as such. From this descends a 1½-inch zinc pipe (*without trap of any kind*) on to the surface of the soil, the total fall being about eleven or twelve feet. This pipe is freely exposed to the air in its entire length, and passes across the roof of an outhouse to reach its destination on the ground. The slops are received on an ordinary garden bed adjoining the north wall of the premises (*i.e.* the wall with southern aspect). This bed was deeply trenched in 1887, and at the bottom was placed all the rubbish that could be found in order to give good bottom drainage. On the surface was arranged, in a straight line and parallel to the wall, and about four feet from it, a few lengths of old zinc guttering which happened to be at hand, through some of which at intervals rough holes were knocked with a pointed coal hammer. The end of the delivery-pipe is laid in this zinc gutter, and thus the slop-water is guided in the right direction. The size of the bed thus roughly 'laid out' for the reception of these slops is three yards broad by nine yards in length, or twenty-seven square yards in all. The soil is a stiff loam lying over brick earth. The whole arrangement was experimental, and was done with the roughest materials, because the writer felt he had his experience to gain, and as his possible failures would have to be paid for by himself, and not by his brother-ratepayers, he moved with caution.

Since this simple contrivance was arranged in 1887, up to the present time (October 1892) it has not been touched, and is still working as well as ever. The ground on which the pipes lie has been tilled and the relative position of the lengths of guttering (some of which have no holes) has been occasionally changed, so as to vary the amount of fluid at particular spots; but beyond this nothing has been done, and good crops of strawberries and spinach have been gathered and plenty of peaches have been plucked from the trees against the walls. There has been no trouble of any kind with this simple arrangement, no smell, no sloppiness; and during the severe winter of 1890-91 the slop-water ran away and soaked into the earth with as much readiness as it does at midsummer. There are no traps and, the fall being considerable, water cannot freeze in the pipe; with the pipes arranged outside the house, it is impossible to have any 'traps,' and that is a distinct gain. The whole arrangement is now covered with creepers and hidden from view.

A similar plan has been adopted with the waste of the fixed bath. A zinc pipe (without traps) has been conducted, by means of a rustic ornamental arch, across the path surrounding the house, and this delivers on to the surface of the ground and pours the water into a gutter scraped in the soil with a hoe, and running through a bed of shrubs (privet, hollies, yew, aucubas and laurels). The gutter wants occasional clearing with a hoe or rake, and the shrubs nearest the water need more pruning than their neighbours, but no trouble has yet arisen with this simple contrivance, which is in the front of the house, a few feet from the parlour window, and between it and the road. The soil is a thirsty one, and when the plug of the bath is pulled up the gutter fills, and two minutes later the thirsty earth has sucked up every particle of the water. This gutter is about fifteen feet in length.

This is a very simple history of a common cockney villa, such as a clerk with 300*l*. or 400*l*. a year might very well occupy, and of which there are thousands in this district. It is rated at 38*l*., and the whole of its bedroom-slops are disposed of in two gutters, one of which is nine and the other five yards long, while the kitchen-slops are given to the shrubs, which make a brave show in consequence in the summer. Of the quarter of an acre of garden only a very few square yards are absolutely needed for refuse utilisation and disposal, and were the material to be used ten or twenty times as great there would be no difficulty. We habitually drink the water from the surface well, which, although hard, runs no risk of contamination. For the fixed bath we have the water of the Grand Junction Company in order to save the necessity of a special cistern and the labour of pumping.

The arrangements which have been described are very simple, but simple as they are they require some attention in the way of clearing away dead leaves or other *débris* which may arrest the current of the water. If anything goes amiss with these simple contrivances it becomes evident instantly, and a rake or garden trowel are the only implements necessary for rectification. To persons who have grass land and who keep a big staff of servants the disposal of house-slops can cause no difficulty. I have been at some pains to show how there is no difficulty in houses of the most modest type.

It is some satisfaction to feel that one lives in a house without sewer or cesspool and with no putrefying collections of filth beneath the surface of the ground.

It is some satisfaction to feel that the refuse of your house is neither a source of annoyance to yourself nor of danger to your neighbour, to feel that there is one house at least in the district (how many more are there?) which sends no filthy water to the river, and the sanitation of which is no expense to the ratepayers and a profit to yourself.

The reader needs to be informed, however, that he has very little liberty in sanitary matters, and that it is dangerous to spend much money upon any system unless it be that particular system which happens to be in fashion and which has received the sanction of the local authorities, and which the local authorities have power to *compel* one to adopt.

The suburban district in which the writer's cottage is situated contains 26,000 inhabitants and 7,000 acres of land, the chief industry is market gardening, and there are very few houses without gardens or ample curtilage. Of course the district fouls the Thames. People have been permitted to send their filth into the Thames, and it is noteworthy that about the worst offender in this respect was a large building estate covered with high-class houses, which is enclosed by gates and fences and has retained its autonomy with regard to its roads. Why this wealthy estate should not have been compelled to deal with its own filth it is difficult to conceive. To have done so would have been easy, because the filth is not mixed with trade refuse of any kind.

But such matters nowadays are never dealt with piecemeal, and the ratepayer gets imbued with the notion that, although he is too lazy or ignorant to deal with his own small amount of refuse satisfactorily, he and his ellow-ratepayers will be able to deal satisfactorily with any amount when their heads are put together and they become a 'board.'

They have only to cast their eyes at London and many places in the neighbourhood to know how insuperable are the difficulties of dealing with sewage on a large scale, how insurmountable are the chemical difficulties, and how well-nigh impossible it is to get the engineering arrangements (which are mainly out of sight) properly carried out by the contractors.

However, it goes without saying that this district adopted a large sewage scheme which involved the taking of pipes in many instances along hundreds of yards of road in order to reach the doors of houses standing in isolated positions and surrounded by acres of market-garden ground. The roads were soon 'up' for the sewers, and eventually miles of pipes were laid and 'sewage works' were erected, the whole costing, so it is said, about 120,000*l.* This figure is doubtful, because the author does not remember ever receiving a detailed statement of the cost in which the ratepayers had been involved.

It may be well to state that the *raison d'être* of these huge works was not the improvement of the public health of the district, but merely the stopping of the pollution of the Thames. The thing was done, and the domestic refuse of 26,000 people was mixed with the trade refuse of the district, including the waste from a big brewery, a dye-works, and a soap works, and this was taken to one spot.

'What are you going to do with your sewage?' I asked of a local magnate one day in the train? 'A process,' was the reply. 'Which?' I asked. 'Oh, I don't know; there's hundreds of 'em.' This answer probably gives a fair idea of the knowledge of the sewage question which is possessed by the average vestryman, who merely sees that a great scheme is 'good for trade,' and especially good for the jerry builder, who invariably follows a line of sewers. The 'process' adopted in this instance was precipitation by means of alum and iron-salts, and the success of it may be judged of by the following extract from the local paper for July 16, 1892 :—

'LOCAL BOARD.

'An ordinary meeting of the Board was held at the Town Hall on Tuesday. . . .

'*The effluent discharge difficulty.*—Mr. J. C. M., a resident of E— Road, protested against the discharge of effluent into the Old River, which flows at the edge of his property. He urged that something should be done immediately to remedy the nuisance, and stated that when he bought the property he was wholly influenced to purchase by the fact of the river being well stocked with fish—affording sport and pleasure. Being very partial to aquatics, he purchased a boat, and built a slip and boathouse, looking forward for some amusement in the future. But since the effluent from the Board's sewer works had been discharged into the river the fish had been driven away or poisoned by the chemicals in the liquor, which was so polluted and disgusting that it made one ill to see it, and the smell was intolerable. He asked the Board to restore the river to its original purity; otherwise he was compelled to dispose of his property under forced sale and remove from the locality, and he would have to call upon the local authority to recoup any loss he may sustain thereby.—Mr. E. said he fully confirmed the statement of the letter. The nuisance was really fearful. Last week the smell was very bad, and the water was extensively polluted. Sometimes that discharge was of a pure and satisfactory nature, but at other times it was directly opposite. Mr. M. having remarked that the matter was receiving careful consideration, the matter dropped.'

The local board had adopted the 'separate' system and a precipitation process aided by chemicals, and already they were in difficulties, and it was an open secret that an inspector from the Local Government Board had condemned their effluent, which was as described in the above cutting.

At this time (summer of 1892) their action had resulted in an increased pollution of the river, and notwithstanding that they had more sewage than they could deal with they were asking for more.

On June 16 the following notice was received :—

Public Health Act, 1875.

LOCAL BOARD.—FORM E.

NOTICE TO DRAIN HOUSE.

To Mr.

The Owner or Occupier of

We, the Local Board for the District of   in the County of   , being the Urban Sanitary Authority for the said district, hereby give you notice that whereas the above-named house, within the said district, is without a drain sufficient for the effectual drainage thereof, we therefore require you, in pursuance of the provisions in that behalf of the Public Health Act, 1875, to make, within the space of one month from the service of this notice upon you, two covered drains from the said house to and emptying into the street sewers, in the mode required by our regulations. The work of making the connections as required by our regulations 2 and 6 will be done by us.

The size of the drains, materials, level and fall to be as stated in our regulations 9 and 10.

Ventilating-pipes, soil-pipes, and other fittings to be provided and fixed in accordance with our regulations.

And we give you further notice that, if the above notice is not complied with, we shall execute the works required and recover the expenses thereof in manner provided by the said Act.

Dated this 16th day of June, 1892.
Town Hall,

Surveyor.

This notice was accompanied by a copy of the regulations, which were as follows :—

Public Health Act, 1875.

LOCAL BOARD.

REGULATIONS FOR HOUSE CONNECTIONS WITH SEWERS AND HOUSE DRAINAGE.

NOTE.—*These Regulations are in addition to the By-laws on this subject—Nos.* 60 *to* 69.

The Main Drainage of the District being constructed on the ' *Separate System,*' *a duplicate system of drains will be required for all houses,* vide *Regulation* 6.

1. *Application for connection.*—The owner or occupier of

any premises desiring to have the same connected with the sewers must make application at the office of the surveyor to the local board on Form A.

The local board will make all connections with the sewers, and lay drain-pipes, &c. beneath the public street up to the houses as follows:—

2. Upon the receipt of such application the surveyor will —so soon as he conveniently can—proceed to lay down 6-in. drain-pipes from the main (foul-water) sewer to the boundary of the applicant's premises, as near the house as the circumstances of the case will permit.

3. *The owner or occupier to pay cost.*—The cost of this work shall be paid in advance by the applicant to the local board, at the surveyor's office, in accordance with a schedule of prices fixed by the board. The amount to be so paid shall in all cases be decided by the surveyor.

4. The expense of keeping in repair the work so done shall be borne by the local board.

But if any of such pipes shall be choked by reason of the placing therein of any substance other than ordinary sewage matter, the owner or occupier of the premises drained by such pipes shall defray the cost of cleaning the same. Where two or more premises are drained by such pipe, the surveyor shall determine by whom and in what proportion the cost of removing any such obstruction shall be paid.

Notwithstanding the payment mentioned in regulation 3, the pipes so provided by the board in consideration of such payments shall remain the property of the board, and no person shall have any claim to them.

5. *Disconnecting chamber.*—A disconnecting chamber shall be provided, the position and size of which shall be fixed by the surveyor.

6. *Rain-water, how disposed of.*—Rain-water will not be permitted to pass into the foul-water sewer, but into a storm-water sewer; exceptions in particular cases may be made by order of the board.

Wherever new or suitable storm-water sewers already

exist, a duplicate system of drains, the one (generally called 'the house-drain') for foul-water and the other for rain-water, must, if the surveyor think necessary, be carried out in such a manner that the least possible amount of rain-water reaches the foul-water sewer from the premises drained.

This duplicate system must also be carried out in all cases where it is contemplated to construct storm-water sewers in the streets, in such a manner that the rain-water can be disconnected from the foul-water drains with which it may have been allowed to be temporarily connected, in as simple and inexpensive a manner as possible.

To obviate in a great measure the necessity of laying a rain-water drain through some premises, the roof gutters of houses should be so arranged as to conduct as much rain-water as possible to the front, to be discharged into the street gutters by open channels, or by not less than 4-in. pipes, into the storm-water sewer in the street, as the surveyor may direct. The work of providing and laying the iron channel-pipe across the footpath, or the 4-in. pipes into the storm-water sewer under the road, will be executed by the surveyor upon the owner or occupier making application on Form A, and paying the sum fixed by the board for such work.

7. *No person shall connect any drain, &c. without giving notice.*—No person shall connect any drain-pipe, soil-pipe, water-closet, urinal, trap, cesspool or other fitting with any drain-pipe communicating or intended to communicate with any sewer, unless he shall have previously given the surveyor one week's notice in writing of his intention so to do. Such notice to be given on Form B.

A plan of the proposed house drainage must be submitted with this notice, such plan to show all necessary particulars and to be drawn to a scale of 20 feet to 1 inch.

NOTE.—A register of sewers and house drains will be kept in the surveyor's office, into which all such plans will be transferred, for future reference.

8. *Nor alter drains.*—No person shall remove or make any alteration in any drain-pipe, soil-pipe, water-closet, urinal,

trap or other fitting communicating with any sewer, unless he shall have previously given the surveyor one week's notice in writing of his intention so to do. Such notice to be given on Form C.

9. *House drains, how to be constructed.*—As far as possible, all house drains shall be laid in straight lines; where changes in direction occur, the same shall be made by bend-pipes, or in manholes, as may be suggested by the surveyor—the pipes to be 6 inches internal diameter for the foul-water drain and not less than 4 inches for the rain-water drain, of good quality stoneware, well glazed, and must be laid with true gradients; the inclination to be 1 inch in a 2-ft. pipe, or as steep as circumstances will conveniently permit, but in no case less than ½ inch in a 2-ft. pipe, and to insure a true and perfectly even inside surface, Hassall's single-lined stoneware pipes, or other equally good pipes, jointed in cement, and to be approved of by the surveyor, shall be used. All pipes to be carefully bedded on the solid ground at least 2 feet below the surface and jointed up so as to be water-tight.

10. *Drain-pipes under buildings.*—So far as possible, no drain-pipe shall pass beneath any building; where such is, however, absolutely necessary, then the pipes shall be of cast iron ⅜-in. thick, coated with Angus Smith's varnish inside and out. Such pipes to be not less than 6 inches internal diameter for foul-water drain, and shall be not less than 4 inches diameter for rain-water drain—to be jointed with lead similar to gas and water mains. No communication whatever will be allowed with the interior of the building.

11. *Ventilating pipes.*—From the top of the disconnecting chamber a main ventilating shaft shall be provided, having a sectional area not less than a 4-in. pipe. This shaft to be fixed to the front or side wall of the building—whichever is the highest and most conveniently situated—and carried to within a foot of the top of the chimney-pots or 3 feet above the eaves, as may be directed by the surveyor. This pipe is to act as an upcast or outlet for sewer-gas. At the other end

of the house drain, and wherever a soil-pipe or w.c. is placed, a pipe or pipes for letting in fresh air must be provided. Such pipes shall be 4 inches internal diameter.

Rain-water stack-pipes shall not be used as ventilators or soil-pipes.

12. *Soil-pipes.*—Soil-pipes shall not be fixed inside any building, but shall be placed outside the walls thereof, and the connection with any closet inside shall be made as short and as straight as possible. Every soil-pipe shall be 4 inches internal diameter, and ventilated by being continued the same diameter above the eaves of the roof, or to such a height as the surveyor may direct. Where such pipe ends near a window a mica-flap inlet cap must be provided. Every soil-pipe shall be either of lead, weighing not less than 6 lb. per square foot, or of cast iron not less than $\frac{3}{10}$-in. thick, properly jointed, its continuation for ventilation to be the same, or of galvanised iron of 20 B.W.G.

13. *Waste-pipes to be disconnected.*—All waste or overflow pipes from sinks, baths, cisterns, safes, &c. must be brought outside the house by the shortest and straightest route, and there discharged in the open air, over a trapped gully, of patterns which may be seen at the surveyor's office.

14. *Drains to be self-cleansing.*—Every drain shall be so arranged as to be self-cleansing, in order that it may be at all times free of deposit. Where this cannot be effected without flushing, proper flushing apparatus shall be provided in the manner directed by the surveyor or inspector acting under his authority.

15. *Stoppage in drains.*—Should any stoppage in any drain occur, information must be given to the surveyor, who will instruct the inspector to ascertain by opening the disconnecting chamber, or by other means, whether the stoppage is in the road, and if it is, to rectify the same forthwith.

16. *Protection against flooding.*—Wherever any premises are proposed to be drained the levels of which are not sufficiently above the invert of the sewers to protect them from the possibility of flooding, the drainage of such places,

if sanctioned at all by the local board, will be at the risk of the owner of the property proposing to carry out such drainage.

Such drains shall be of iron and provided with a reflux valve, and also a screw-down stop-valve.

And in localities subject to flooding from Thames water, every closet, sink or other inlet to the sewers shall be fixed at or above the level of 18.50 above Ordnance datum. And to prevent the flooding of the sewers by percolation into the house drains, these drains up to the level of Trinity High-water mark, or 12.50 above Ordnance datum, shall be laid with iron pipes, jointed with lead, or Hassall's double-lined pipes, or other approved quality, jointed with cement, the same as the sewers.

17. *Water-closet supply cisterns, seats, &c.*—All water-closet pans and fittings shall be of the simplest description. No container, D trap, or other similar fitting will be allowed under any pan.

Every water-closet that is not a valve-closet shall be supplied by not less than a two-gallon waste-preventer cistern of approved pattern, fitted immediately over the closet, the down-pipe from which shall be $1\frac{1}{4}$ inch internal diameter, and shall be fitted as straight as possible; the bottom of the cistern shall be at least 5 feet above the closet-seat.

All valve-closets shall be of approved pattern, with after-flush, and shall be supplied by a 1-in. pipe from an independent cistern to that which supplies the house.

The seats of all w.c.s shall be so constructed as to be easily removable, in order that every part of the closet may be inspected with facility.

The trap of every indoor closet shall be provided with a 2-in. ventilating-pipe, from the top of the trap to the soil-pipe ventilator outside the building, above the highest inlet into the same, or independent of the soil-pipe.

18. *Work to be inspected before covered up.*—No builder, plumber, or other workman shall be allowed to do any work in connection with the drainage of any premises unless he

agrees to conform to these regulations; and all work shall be executed in every respect in accordance with these regulations. All such work shall be inspected by the surveyor or inspector appointed for that purpose, and every facility shall be afforded for such inspection. No underground or enclosed work shall on any account be covered up or concealed from view until the same shall have been duly inspected and passed by the surveyor or inspector.

19. *Defective drains, &c.*—Any drain-pipe, soil-pipe, trap, water-closet, urinal, sink or other fitting, laid, used, or constructed otherwise than in accordance with these regulations and with the provisions of the Public Health Act, or which shall in the opinion of the surveyor be of defective quality, shall upon notice in writing from the surveyor be removed or repaired in the manner determined by the surveyor.

20. *Notices, &c.*—All notices and applications required by these regulations are to be made upon the printed forms to be obtained at the surveyor's office. The surveyor's approval to all plans submitted with such notices will be given on Form D.

21. *Cesspool filled up.*—Upon the completion of the connections between any premises and the sewers, the owner or occupier of the said premises shall—whenever required by notice so to do (Form E)—construct a proper water-closet or closets in accordance with the regulations, and immediately thereafter empty, cleanse, and fill in all cesspools and other receptacles for sewage matter upon the premises, to the satisfaction of the inspector.

22. *Ventilators and drains, &c. to be kept open.*—All openings for ventilation made in accordance with these regulations or by order of the surveyor shall at all times be kept open and perfectly free from obstruction.

Every occupier shall at all times see that all drains upon his premises, and that all traps and other fittings, are at all times in good order, clean, and free from obstruction.

To prevent such things as rags, cabbage leaves, hair,

pieces of soap, &c. &c., from passing into the drains, all openings to the drains must be protected by proper gratings.

Surveyor.

Approved and adopted by the Local Board this 11th day of February, 1890.

Clerk. *Chairman.*

The attention of the ratepayers is especially called to the following sections of the Public Health Act of 1875 :—

Sec. XXI.—The owner or occupier of any premises within the district of a local authority shall be entitled to cause his drains to empty into the sewers of that authority, on condition of his giving such notice as may be required by that authority of his intention so to do, and of complying with the regulations of that authority, in respect of the mode in which the communications between such drains and sewers are to be made, and subject to the control of any person who may be appointed by that authority to superintend the making of such communications.

Any person causing a drain to empty into a sewer of a local authority without complying with the provisions of this section shall be liable to a penalty not exceeding twenty pounds; and the local authority may close any communication between a drain and a sewer made in contravention to this section, and may recover in a summary manner from the person so offending any expenses incurred by them under this section.

Sec. XXIII.—Where any house within the district of a local authority is without a drain sufficient for effectual drainage, the local authority shall, by written notice, require the owner or occupier of such a house, within a reasonable time therein specified, to make a covered drain, or drains, emptying into any sewer which the local authority are entitled to use, and which is not more than one hundred feet from the site of such house; but if no such means of drainage are within that distance, then emptying into such covered cesspool or other place, not being under any house, as the local authority direct, and the local suthority may require any such drain or drains to be of such materials and size, and to be laid at such level, and with such fall, as on the report of their surveyor may appear to them to be necessary.

If such notice is not complied with, the local authority may, after the expiration of the time specified in the notice, do the work required, and may recover in a summary manner the expenses incurred by them in so doing from the owner, or may by order declare the same to be private improvement expenses.

These regulations show conclusively how arbitrary is the power possessed by any local authority, and how great is the expense to the householder (in addition to the increase of rates) of complying with the requirements. These requirements also amply account for the popularity of 'sewage schemes' among builders, plumbers, the proprietors of patent pipes or varnishes, and also among water companies, for be it observed that the adoption of water-closets makes one dependent upon a water company for cleanliness. The whole regulations are remarkable for their precise directions for wasting rain-water, and contain no single word as to its storage. The document is also remarkable in so much as an earth-closet is not even mentioned as a possible alternative for the system proposed; notwithstanding that every E.C. erected would save the Thames from pollution and tend to lessen the difficulties into which the ratepayers have been run by the 'board.'

The above documents have been printed in order to show the householder what a 'sewage scheme' means, and how with its adoption the Englishman's house ceases to be his castle and becomes a mere profitable plaything for patentees. The name of the board has been suppressed because the writer is well aware that its actions are largely controlled by higher authorities; and the actions of this particular board do not differ materially from those of others. The facts are given merely as types of what is common. The writer has no complaint against individuals, and he gratefully admits that, except in sending a peremptory order to foul the Thames, the action of the board towards himself has been reasonable and considerate. In answer to the above circular the board was informed of the arrangements which have been described, an official visited the premises, and having assured himself that we neither polluted the Thames nor annoyed our neighbours, and that we were not poisoning ourselves, he appeared to be satisfied, and no further action has been taken.

This cottage is probably the only one of the group of fifty houses to which it belongs that is without sewer or cesspool, and positively does not endanger the Thames. It would, however, be quite feasible for all the others to adopt similar measures, and many probably would do so but for the compulsory power of the board, which makes occupiers chary of spending money upon arrangements which the board has the legal right to destroy, and then further compel the householder to spend money not as he fancies, but as 'We the Board' order.

Then, again, people think that, as they pay the rates, and as the sewers are in any case an expense to them, they may as well connect; and they are further stimulated by the thought that obedience to the board is the readiest road to peace and quiet, and as for the 'effluent,' that is an affair for the board, the Thames Conservancy, and the Local Government Board to settle between them.

It is not (but it ought to be) a matter which concerns the individual.

The above little history has shown how a great sewage scheme was undertaken to save the Thames from pollution, and how the pollution was worse than ever when the scheme was finished. It also shows that at a time when the 'board' was in trouble about its effluent, and was receiving more sewage than could be satisfactorily treated by its existing plant, it was nevertheless serving peremptory notices upon householders which could have but one effect, viz. the increase of its own difficulties. It need hardly be said that the usual complaints are being made about the smells from sewer gratings, &c., and the inevitable question of 'ventilation' has cropped up. Indeed it may safely be predicted that the sewage difficulties of this district, so far from being ended, have only just begun, notwithstanding an expenditure (both public and private) which cannot be far short of a quarter of a million of money, or nearly 10*l.* per head of population.

The poor are apt to fancy, when they see the streets and roads blocked by 'works,' that it is good for trade and they do not stop to think that possibly money which might go into their pockets is being sent out of the district for the purchase of patent pipes, machinery and chemicals, and that the work itself is largely executed by labourers imported by the contractor. These schemes are thoroughly bad for the labouring classes, because they seek to do by mechanical means that which can only be done efficiently piecemeal and by the aid of hand labour. Such schemes starve the ground and cause less money to be spent in cultivating the soil and in harvesting crops.

A quarter of a million of money (the sum probably spent by the 26,000 persons of the district) at 3 per cent. will yield 7,500*l.* per annum, or enough to pay good weekly wages *all the year round* to 150 extra scavengers, and the wages of the staff, the cost of maintenance and the sums paid for chemicals would probably support fifty additional scavengers. There can be no doubt that, from the point of view of employment for the poor and the provision of work all the year round, these big schemes are thoroughly bad. And, further, there is no doubt that, if the sums spent on pipes and machinery (public and private) had been sunk for the provision of extra labour, the Thames would have been saved from pollution, there would have been no annoyance, and a profit would have been made. Most certainly the poor labourer and those who are anxious to provide constant employment for the working classes should in no case support big sewage schemes, which rob them of work. If the sums raised by 'rates' be spent in the district not much harm will result, and rates spent in labour will probably save the 'poor-rate.' But if huge sums be spent upon imported machinery which is erected by labour which is also often imported, and if such machinery and plant lead to the employment of less local labour and the starving of the soil of the district, then it is evident that big sewage schemes, which are bad for everyone (except the landowner, jerry-builder, and water shareholder), are especially bad for the poor. Even if the scheme be entirely carried out by local labour and home-manufactured materials, the gain is only transient, and we have to consider the relative advantages of spasmodic *versus* permanent employment.

Refuse, if properly used, is a source of food and wages, but if improperly used it merely leads to waste and starvation. That is a fact which the poor man has to bear in mind when he votes at municipal elections.

It will be profitable to consider the financial aspect a

little more in detail. The rateable value of the writer's cottage is, as has been stated, 38*l*., and for the first half-year of residence (1887) the rates were: district rate, 1*l*. 18*s*.; poor-rate, 1*l*. 18*s*.; burial-rate, 6*s*. 4*d*.; or a total of 4*l*. 2*s*. 4*d*. (8*l*. 4*s*. 8*d*. for the year).

Since then the rates have gradually crept up, and for the last *half*-year (1892) the amount was: district rate, 3*l*. 16*s*.; poor-rate, 2*l*. 4*s*. 4*d*.; burial-rate, 6*s*. 4*d*.; giving a total of 6*l*. 6*s*. 8*d*. (12*l*. 13*s*. 4*d*. for the year). Thus in five years the rates have risen 53 per cent. The last demand note issued by the District Board contains some information concerning the details of expenditure, &c. which is worth recording :—

ESTIMATED RECEIPTS AND EXPENDITURE
For the Half-year ending March 25, 1893

LIABILITIES

	£	*s.*	*d.*	£	*s.*	*d.*	£	*s.*	*d.*
Salaries				892	6	0			
Establishment				250	0	0			
Works Committee, Estimate:									
Lighting and highways	7,400	0	0						
Sewage disposal works	1,054	0	0						
				8,454	0	0			
Fire Brigade				600	0	0			
Hospital				150	0	0			
Loans, interest, and repayments				3,013	8	4			
Sundries (including creditors)				2,600	0	0			
							15,959	14	4

ASSETS

	£	*s.*	*d.*	£	*s.*	*d.*
Balances				1,398	6	0
Arrears of rate				860	0	0
County Council (main roads)				3,287	10	0
Miscellaneous				652	11	8
				5,198	7	8
To be provided				10,761	6	8

The rate demanded is two shillings in the pound.

It is to be noticed that the County Council provides 3,287*l.*, an asset which was non-existent in 1887, when the rate levied was a shilling only. If the District Board had to collect this sum for themselves it would raise the demand to 2*s.* 7½*d.* or 2*s.* 8*d.* in the pound. This windfall from the County Council might, but for the large expenditure on sewers, have reduced the shilling rate of 1887 to one of fivepence only, and instead of a demand for 3*l.* 16*s.* for the half-year, one of 15*s.* 10*d.* only would have had to be made. As it is, the County Council money merely saves the board from having to demand 4*l.* 18*s.* 2*d.* instead of 3*l.* 16*s.*, and serves mischievously to mask their extravagance. We have said that the total yearly rates levied on a 38*l.* house amount to 12*l.* 13*s.* 4*d.* But if we reckon in the County Council money (which in fairness ought to be reckoned), and if to this be added the inhabited house duty (1*l.* 2*s.* 6*d.*), income-tax (1*l.* 2*s.* 6*d.*), and tithe (3*s.* 10*d.*), it will be seen that the real charges on this house amount to 17*l.* 6*s.* 6*d.* Notwithstanding this high figure, it is said that we ought to be thankful because we have some well-endowed schools, and are saved a school board rate, which (judging by London experience) would very soon run up the charges on the house to 19*l.* or 20*l.*

Again, it must be remembered that with a compulsory system of sewerage (and compulsory w.c.) there is practically a compulsory water-rate which (with a 38*l.* house) amounts to 2*l.* irrespective of the amount of water used.

We have thus arrived at the conclusion that, although the actual charges in this particular instance amount to only 17*l.* 6*s.* 6*d.* (reckoning the County Council money), we ought nevertheless to be thankful, and that the average villa holder in a London suburb may think himself extremely lucky if the obligatory charges on his house amount to less than 50*l*. per cent. of the rental. The clerk on 400*l*. a year who takes a 40*l*. villa in the suburbs must remember that the total outgoing for his house will really amount to 60*l*., and that in these days he will not only have to pay income-tax, but a special local taxation as well, which will not improbably amount to 5 per cent. of his total income.

Looking again at the details of the board's demand note, we find that the interest and repayment of loans amount to 3,013*l*. 8*s*. 4*d*., or 6,026*l*. 16*s*. 8*d*. for the year. The details of the loans are not given, but the sum paid for interest looks as if the indebtedness of the board was about 150,000*l*., of which over 100,000*l*. is for sewerage works. If a private individual borrows 100,000*l*. he expects to be able to apply the sum profitably and earn something more than the interest. This is not the case with sewerage works, the money for which is spent in erecting costly machinery for wasting money without the possibility of any return whatever. Thus we are not only spending some 4,000*l*. a year as interest for our pipes and works, but the sewage disposal is already entailing an annual cost of 2,108*l*. (1,054*l*. for the half-year). Moreover, it is certain that we are not at the end of our troubles, for not only is there a talk of another loan for filter-beds with the object of cleansing the mess which has resulted up to the present, but the whole plant being liable to wear and tear the cost of maintenance is sure to be far greater in the future than it is at present.

Lastly, as to the 1,054*l*. for 'sewage' disposal works, we have the following items :—

Wages . . . . 364*l*.
Ferozone and alum . . . . . . 240*l*.
Coals . . . . 250*l*.
Lime . . . . 60*l*.
Repairs and maintenance. . . . . 100*l*.
Sundries . . . . 60*l*.

These figures are for half a year only.

The only remarks which need be made on these items are: (1) That the expenditure on iron salts (ferozone) must make the sludge and effluent very dangerous for any agricultural purpose, and (2) that the amount for maintenance and repairs will not probably be so small in the future.

We are just now (December 1892) being threatened with an increased poor-rate to find work for 'the unemployed.' Time out of mind the work of scavenging has been done by the class just above pauperism. We are spending 6,000*l*. a year in a futile attempt to get our scavenging done automatically; we are robbing the lowest class of what may be regarded as their legitimate employment, and as a natural consequence we are being called upon to keep them.

WHAT SHOULD BE DONE?

In the foregoing chapters the author has given abundant reasons to prove that the methods at present adopted in our treatment of organic refuse are unscientific.

If our methods be bad in a scientific point, they must also be bad both morally and economically, and much has been said in support of both these latter positions.

It is easy to condemn, but it is not so easy to remedy. If a start be made on an unsound scientific basis, remedy is impossible, and I believe that for

London, and the other big towns which have imitated her, there is no remedy. They must continue to blunder on as they have blundered in the past, and endeavour by huge expenditure and a merciless taxation of the householder to counteract the evils in connection with polluted rivers, overcrowding, epidemics, and dwindling water-supply — evils which have increased, and will increase, because the hygienic arrangements of towns are more influenced by considerations of 'business' and immediate profit than by scientific considerations and a wise thoughtfulness for the future.

If the big towns merely serve as a warning to the country, and a standing example of 'How not to do it,' they will serve a very useful purpose.

Although *remedy* is hardly to be thought of for huge places which have got into a sanitary *impasse*, *prevention* is easy in places which still retain a fairly rural character.

The writer believes that an equitable adjustment of sanitary rates would prove sufficient to prevent the primary evils of unscientific sewage treatment and its secondary evil of overcrowding.

Our sanitary arrangements have a basis of pure socialism, or rather one would say of lop-sided socialism, for not only are the rich taxed to provide sewers and water for the poor, but the sanitary saint is taxed for the salvation of the sanitary sinner, to whom no punishment is ever meted out.

1. One of the reforms most urgently needed *is the supply of water by meter*. The possession of water under pressure is a priceless boon if it be not abused, and the only way to stop the abuse of water is to charge for it in proportion to the quantity used. If water were supplied by meter, and if water companies were made to adopt a

sliding scale by which the rate of dividend and the price of water bear an inverse ratio to each other, many of our sanitary troubles would be at an end. All waste of water would cease, and with it the volume of sewage would decrease, and scientific methods of treating refuse would be adopted if it were found economical to adopt them.

It would, of course, be the house landlord's duty to provide an adequate supply of water to weekly property, and it would be the duty of the sanitary inspector to see that such property was kept clean. Poor men and women are constantly being fined for such a trivial offence as not sending their children to school, and one hopes that eventually some notice may be taken of sanitary sins which endanger the health of the neighbours. If this cannot be done, let us abolish our sanitary inspectors as being of no use.

The charging for water by rateable value is most inequitable, and now that the London County Council is claiming to arbitrarily fix the rateable value of our dwellings without reference to the sum paid for rent this iniquity is likely to increase. The writer's London house has an average of four inhabitants, and yet more is being now paid for water than was the case a few years ago when the same house had nine inhabitants.

2. No cesspool or underground sewer should be permitted to be constructed, even on private property, without license, because such underground collections of filth (whether in sewer or cesspool) have been shown to be capable of contaminating wells at a distance which is almost illimitable, and are a distinct danger to the community.

3. All water-closets which discharge into a public sewer should be taxed,

4. Those who make no contributions of foul water to a public foul-water sewer should not be called upon to pay for its construction or maintenance, provided that their sanitary arrangements be satisfactory, and not likely to cause annoyance or danger to neighbours.

5. All sewers and sewage works should be constructed and maintained entirely out of rates levied on the ground landlords, because it is they, and they only, who make a profit out of that overcrowding of houses on space which is only rendered possible by sewage schemes, and which is too often the main reason for their inception. If the owner of a building estate choose to act independently and deal with the organic refuse of his estate, and if no foul water escape from his estate into a public stream, then, of course, he should not be called upon to pay for the construction of sewers, which are not only of no use, but a positive danger and annoyance to him.

6. The pollution of rivers ought not to be tolerated, and the Act for its prevention most certainly ought to be enforced against individuals.

7. It is, of course, imperative that all manufactories should be compelled to deal with their waste products, and should not under any circumstances be allowed to discharge factory waste of any kind into public sewers or streams. Near the writer's suburban cottage are a big brewery and a soapworks, both recently converted into 'Limited Companies,' and both paying good dividends. Why these two wealthy companies should not be compelled to keep their effluent out of the sewers and to filter it at home instead of having it done at the expense of the ratepayers is a mystery. The wealthy manufacturer who fouls rivers should be dealt with rigorously and mercilessly.

The question of manufacturing waste and sewage is a most important one. It is the manufacturers' sewage of unknown and very variable composition which puts great difficulties in the way of public authorities who are compelled to undertake the purification of town refuse. It is often poisonous, and generally, from an agricultural point of view, most dangerous, and effectually destroys the manurial value of sewage or sludge. The refuse and sewage of each trade are peculiar to themselves and need peculiar treatments, and no sanitary reform is more urgently needed than the compelling manufacturers to deal with their waste products and send a pure effluent to the rivers. What right have these wealthy manufacturers to be literally supported out of the rates like paupers?

If manufacturers were compelled to deal with their own refuse and waste products, the factories would almost certainly be compelled to occupy more space than they do at present, and this rule, which seems equitable and reasonable, would have the effect not only of saving the ratepayer's pocket, but of checking that excessive overcrowding which is such a danger to the health of towns.

The populations of our big towns are kept alive by free trade, and the removal of duties on imported food has lessened the profits of the agricultural population. Protection will never be tolerated again until the small agricultural holders outnumber the artisans, and that day is far distant. It may be said that the agricultural interests of the country have been largely sacrificed for the benefit of the manufacturer, and in exchange for this the agriculturist has a right to ask that the manufacturer shall neither be permitted to foul the stream from which his cattle have to drink, and which when pure is a source of profit to him in many ways, nor be allowed to destroy the manurial value of town sewage by allowing strong chemicals to find their way into the sewers.

The country has made great sacrifices for the sake of the towns, and gets nothing in exchange but foul water and smoke.

The writer is not inclined to recommend any legislation, of which we have had far too much already, and which has caused precipitous action of a disastrous and ruinous kind in many parts of the country. What is wanted is more freedom of action, more encouragement for those who are striving to do well, and are a cause neither of annoyance nor expense to their neighbours, more rigorous treatment of the evildoer.

CHAPTER XII

BURIAL

'All go unto one place; all are of the dust, and all turn to dust again.'—*Eccles.* iii. 20.

THE question of burial is but a part of the great question which we have been considering in this volume, viz. the power of the living earth to deal with organic refuse.

A dead body, like other forms of organic refuse, becomes a prey to lower forms of life, and ultimately serves as food for plants. If we try to stop or hinder a natural process, the dead body becomes a nuisance and a danger. Nature's laws cannot be revoked. They are eternal. She can bide her time, and sooner or later her beneficent work is accomplished, in spite of the powerful opposition of self-interest, ignorance and superstition.

What with shells, leaden coffins, oak cases, silver fittings, brick vaults and vulgar tombstones, burial is a most expensive affair, and although we profess to believe that there is no rank in the tomb, a visit to the nearest cemetery will prove that in death arrogant ostentation reaches its highest pitch.

Sometimes a railway cuts rudely through a cemetery and presents us with a section of it, a section which reveals in all its ghastliness the horrors of modern Christian burial.

Mr. Seymour Haden has done good service by drawing public attention in forcible and eloquent language to the disgusting nature of our present mode of burial. If a body be laid in the earth, and if the earth come into actual contact with the body, its gradual resolution into nitrates and humus is effected quickly and without offence, and if the surface of the grave be planted with rapidly growing trees and shrubs, the purification of the soil and of the air above it is insured. The roots find their way to the organic matter, and in due time green leaves and fragrant blossoms tell us of the wondrous process which has been going on below.

The power of the earth to appropriate to itself all decaying organic matter is most astonishing. During the time of the plague thousands of bodies were shot into pits and covered up, and no evil has ever been attributed to this process. Those who walk up Regent Street little think that, as they pass the end of Beak Street, they are within a few yards of one of the plague pits of 1665, into which many thousands of the victims of the epidemic were cast.

If we wish a dead body to be decently turned again to the condition of inorganic matter, we must actually commit the body 'to the earth,' and we must take care that the expressive language of our burial service is not turned to a mockery and a mere figure of speech. At the most a wickerwork coffin must be used, or a coffin so contrived that it can be withdrawn and allow the soil to come into actual contact with the corpse. Lead and oak are not to be thought of. It is the fashion to bury bodies very deeply, but it is evident that the resolution of the body must proceed most quickly in the upper layers of the soil—*i.e.* in the living earth where the aerobic saprophytes will quickly nitrify it.

Brick graves should be abolished, and it should be

BURIAL

made illegal to bury more than one body in the same grave. Each body should be completely surrounded by earth.

Next to impervious coffins and brick vaults the worst thing is the placing of a slab of impervious stone on the surface, thereby preventing vegetation, and preventing also to some extent the access of moisture and air.

If we could all be reasonable on this point, the cemeteries of one generation might become the recreation-grounds of the next. Cemeteries should be on the outskirts of towns, and the builder of every new house should pay a fixed sum into the cemetery fund; for every house in the course of time must provide occupants for the cemetery.

The family of the deceased might, if they wished, have the lease of a plot of ground (say 9 feet by 3 feet) in which to dig the grave. This plot of ground should revert to the community at the end of ten or twenty years, when the poignancy of grief would have abated, or possibly the deceased have been forgotten. The grave should be planted with trees and shrubs most suitable to the soil, and big tombstones should be prohibited. In order to give some scope for memorials, a cloister might be erected for the reception of monuments and inscriptions.

Three or four pounds is usually paid for a grave in a public cemetery, so that for the twenty-seven square feet necessary for the decent interment of an adult as much as three pounds may be asked. At this rate 1,613 persons might be buried with ease and without offence in an acre of ground, and the fees for land would amount to 4,839*l*. Counting children, we may assume that, by adopting this leaseholding plan, 2,000 bodies could be interred in an acre. If this were done in London, some forty acres of land would be necessary for the interment of the yearly dead, which amount to some 80,000. It is evident, from the calculation of the burial fees, that a very handsome price might easily be afforded for the land.

At the end of twenty-five years this land should revert to the community, and if this course were pursued there would be an end of our trouble in respect of open spaces in crowded cities, for every man in dying would benefit his neighbours by giving them the reversion of twenty-seven square feet at a cost of 3l. on his estate. This co-operation of the dead for the benefit of the living would solve many difficult problems, would do something to arrest overcrowding, and would establish a certain sure relation between buildings and open spaces, which is very desirable. Such a course as this would not outrage any of our feelings; it would not prevent any manifestation of respect to the deceased; it would not interfere with any religious forms; and it may reasonably be hoped that those who are shocked at the idea of an innocent child playing over the spot where an interment took place a generation back are at least a diminishing minority.

It may be hoped, too, that at least the custom will some day come into vogue of planting trees and not tombstones over the spot where those we love best repose. A tree, as a memorial, improves with age, while the mouldering, moss-grown, untidy tombstone merely teaches us that 'the dead know not anything, neither have they any more a reward, for the memory of them is forgotten.'

There are some who advocate cremation in place of burial in the earth, but I hold that cremation is not necessary; that it is almost hopeless to expect a general assent to the practice; that it is wasteful; and that it would often defeat the ends of justice. Burial in the earth, if decently and properly conducted, is a safe and healthy practice.

If cemeteries be properly controlled, and if burial be conducted upon scientific lines, it is inconceivable that any harm can result.

In a paper communicated to Section IX. of the Seventh International Congress of Hygiene and Demography (London 1891), entitled 'L'Assainissement des Cimetières,' MM. Brouardel, Du Mesnil and Ogier state:—

1. '... que l'atmosphère des cimetières actuel est pure de tout produit gaseux délétère, de tous les éléments figurés nocifs.

2. 'Que dans les cimetières actuels le sol ne renferme que de l'acide carbonique en grande quantité à l'exclusion de tout autre gaz en quantités appréciables.

3. 'Que la décomposition des cadavres confiés a la terre serait vraisemblablement activée par le drainage du sol.'

And in a second report they further state:—

1. 'Que plus le cadavre est en contact avec l'air par le fait de la perméabilité du sol, de la porosité de la bière, plus sa destruction est rapide et complète; la présence de l'air favorisant l'éclosion de ces êtres inférieurs, *de ces travailleurs de la mort* qui sont les agents les plus actifs de la destruction des cadavres.

2. 'Que toute substance mise dans la bière pour en assurer l'étanchéité—sciure de bois mélangée ou non de substances antiseptiques ou simplement aromatiques, poussière de charbon, feuilles de caoutchouc, de carton bitumé doublant la bière—retardent dans une proportion considérable la destruction des cadavres.

'Toutes ces matières s'opposent au développement des

animalcules la plupart aérobies qui paraissent être les agents les plus actifs de la destruction des cadavres.

3. 'Que l'inhumation dans un sol humide ou imperméable retarde considérablement la putréfaction.'

This extract seems to show that in burial as in other things we have only to follow nature in order to attain our end, and that the unwholesomeness of the well-managed cemetery is illusory.

The advocates of cremation use as their stock argument the danger to the health of the living that is caused by the burial of the dead. The germs of zymotic disease, they urge, are thus kept alive and find their way to the wells and watercourses. That the germs of zymotic disease escaping from the bodies of the yet living sick do undoubtedly leak from sewers and cesspools to our water-sources and occasion epidemics of disease there can be no doubt, but the author is not aware of any similar epidemics which have been traced to cemeteries, not even where the burial is carried out with an absolute defiance of all scientific teaching. Those who have read the foregoing chapters will be ready to believe that, if a body be placed in the upper layers of the soil, the earth will serve as an absolutely perfect filter, and that nothing but soluble nitrates and other mineral salts can possibly be carried in the rain as it percolates from the surface to the wells or springs. As for the zymotic germs in the body, they become the prey of saprophytic fungi and absolutely disappear.

The author is a very strong advocate for burial upon scientific lines, and he believes that if a body be committed to the earth with due attention to those details which will insure its humification nothing but good can result. The humus is the most perfect filter for bacteria which is known, and it is only when we prick a hole in our filter and conduct infected liquids to underground and leaking receptacles that our wells are in danger.

Granted that in the dead body, as in the excremental and other matters escaping from the sick, there may be living particles capable of carrying infection, we should expect to find that those most associated with the dead and with excremental matters would give some indication of this danger if it be practically important.

As a matter of fact, the three healthiest classes are (1) the clergy who habitually live by the side of a graveyard and get their water from a local well, (2) market-gardeners who are constantly dealing with enormous quantities of manure, a large part of which must be theoretically capable of conveying infection, and (3) farm labourers who run the risks of market-gardeners but in a less degree.

In considering the question of Burial v. Cremation we must be careful to separate essentials from non-essentials. The pomps and vanities which have grown up round the simple act of interment have become intolerable, and there can be no doubt that the certain amount of popularity which cremation enjoys is due to the fact that it affords a ready means of escape from these oppressive conventionalities.

There can be no doubt that, if we wish to maintain the fertility of a country, all dead organic matter which emanates from the soil must be returned to it.

There are at present enormous tracts of fair agricultural land which can be bought for 20l. or even 10l. per acre, so that it is idle to talk as if the procuring of land for cemeteries were a serious difficulty.

The cremation of a body involves a needless dissipation of energy. Not only has the fuel used for cremation to be paid for, but the air is fouled with the products of combustion, and the inevitable destiny of those products (*i.e.* to become food for plants) hindered to an extent which we cannot estimate.

To regard the dead body as a fertiliser may shock the sensibilities of some, but looked at philosophically this point of view is inevitable, for, even when cremation is practised, the products of combustion must ultimately become the food of plants, somewhere and somehow. Why not go with Nature instead of against her? Why not employ the economic process of inhumation instead of the extravagant and thriftless process of cremation?

There is only one way of keeping the soil pure and wholesome, and that is to make it bring forth. In time of war or famine there could be no objection on sanitary grounds to growing any ordinary crop in a cemetery, but we shall all agree that the most suitable crops would be flowers and flowering shrubs, and it should be an object to make our cemeteries as beautiful as possible.

One is at a loss to understand what is meant when advocates of cremation talk of inhumation as a topic too ghastly to be mentioned. There is no offence whatever, and from the æsthetic point of view the advantages are altogether on the side of scientific burial.

The products of cemeteries must have a definite value, and would provide a large amount of firewood, and this secondary cremation, by which the dead provide warmth for the living, is a cremation to which nobody can object.

A body buried in the earth nitrifies, and the rain, percolating through it and becoming charged with harmless necessary nitrates, will bring fertility wherever it may trickle. The leaves of trees planted in cemeteries, and falling in the autumn in the cemetery itself or on adjacent ground, must appreciably increase the fertility of the soil.

At present we are busy in burning all the combustible refuse of our towns, and by gross mismanagement our excremental matters have become a danger instead of a blessing to the soil. Agricultural depression is, one can have no doubt, in part at least attributable to this cause, and now comes the proposal to cremate the dead, which will foul the air and starve the soil still more.

If the interment of the dead be conducted upon scientific lines, and if the cemeteries be made to produce trees and shrubs, the soil may be used for interment (after an interval) over and over again, and there can be no doubt that its ability to nitrify and humify the body would progressively increase with each repetition of the act.

The process of cremation involves, as I have said, a dissipation of energy, and the same remark is equally applicable when any organic matter which is capable of nitrifying and humifying is destroyed by fire. Fire brings destruction, but organic refuse, when buried in the earth, is started upon a round of creation which (we cannot help it) freshens the air, feeds the hungry, clothes the naked, warms the shivering poor, and finds labour for 'the unemployed.' In cities we are so closely packed that it is no longer feasible to put organic matter to its proper use. Your true cockney browses on imported corn and tinned food, and is necessarily ignorant that organic refuse has any legitimate destiny except to foul the Thames or feed somebody's patent 'destructor.' He cries to be relieved of his organic refuse, which is to him merely a nuisance to be got rid of. He has entered upon a policy of destruction which is unscientific and thriftless, and the proposal to cremate the dead is merely a part of a policy which will eventually do for England what a similar policy in ancient Rome did for the Roman Campagna.

If a piece of ground be set aside for a cemetery, it can, if treated scientifically, be used for ever, for, as has been stated, the power of the earth to deal efficiently with organic matter tends to increase with use. There is great need of accurate knowledge as to the time which dead bodies take to humify, and most certainly we need a patient scientific inquiry into the matter. Professor Flower has stated that when he prepared the skeleton of a whale by burying it he found that in two years from the time of burial all the flesh had disappeared, and only the bones of the whale were left, and there can be no doubt that, if burial be scientifically carried out, a very few years would suffice to humify the body, and the ground would be ready for the reception of a second.

If burial were carried out scientifically, it is to me quite inconceivable that any danger could arise from the germs of zymotic disease, either in the air above the cemetery or in the wells beneath it, and the idea of the earth becoming 'foul' under such conditions is not tenable. In order to make assurance doubly sure, we should probably do well to follow the example of our ancestors in regarding our cemeteries as 'sacred spots' dedicated for ever to the service of the dead. The sanitary aspects of consecration would be quite worth dwelling upon, because, if this ceremony be not a mere mockery, churchyards and cemeteries should remain for ever as open spaces to freshen the air with verdure and separate the living in our overcrowded cities. If we allow the railway contractor to cut through our cemeteries, and if we are willing to buy up city churches and their surroundings in the belief that every bit of ground that does not carry a pile of offices is wasted, and if the clergy are willing to barter away these 'consecrated spots,' we must not grumble if the evils of overcrowding, physical, moral, and political, become rather more than we can conveniently grapple with.

To sum up the merits of cremation and inhumation we must admit that, with regard to ceremonies to be observed, there need be no difference, and that in both it is equally necessary to provide for the transport of the body and a suitable preparation of it, sufficient to allow of its being decently placed in a hole, and whether this hole take the form of a furnace or a grave makes no difference.

There can be no doubt that burial is the shorter process, for to dig a hole having a capacity of 36 cubic feet, to lay a body in it, and cover it up again, would not take many minutes.

Next, let us look at the financial aspects of the question.

If we allow for each adult interment the space of 6 feet by 2 feet, then 3,630 adult bodies could be placed in an acre of ground, and there can be no doubt that, if it were necessary, it would be possible by a little management to utilise every square inch of an allotted space without superimposing one body on another.

It is stated in Chambers's 'Encyclopædia' that the cremation of a body at Woking occupies about an hour and a half, and takes about seven shillings' worth of fuel. It is evident that the cost *for fuel only* of cremating the number of adults capable of being buried in an acre of ground would be 1,210 guineas (1,270*l.* 10*s.*), and this represents the price which might be paid for land for burial purposes before cremation could in comparison with earth burial be considered economical.

Any land which is in a condition to produce ordinary

farm crops may be regarded as suitable for burial, and one need not insist that land used for burial must be properly drained, &c.

The writer calls to mind a small parcel of land within five minutes' drive of an important railway junction between sixty and seventy miles from London, which recently cost 33*l.* per acre. If this ground were filled with one layer of adult bodies, each body having $6 \times 2 = 12$ square feet, then each body would occupy almost exactly two pennyworth of ground. There are hundreds and thousands of acres of such land to be had for similar or lower prices, and it is not necessary to cry out about the difficulty of obtaining land for burial purposes. It is needless to urge that ground used for burial must become exceedingly fertile, and should be correspondingly wholesome, profitable and beautiful.

From the financial point of view there can be no doubt that burial is cheaper than cremation. The burning of a body plus seven shillings' worth of fuel must foul the air. Granting that smoke and offensive odours are both done away with by using proper furnaces, the ultimate products of combustion must render an enormous volume of air noxious and irrespirable. The atmosphere of this country is sufficiently fouled as it is without needlessly adding to it.

The ground in which a body is scientifically buried would be soon covered with herbage and green leaves, which serve to freshen the air instead of fouling it.

The act of burial necessitates the retention of open spaces near cities, a fact which must be allowed to more than counterbalance the dangers of burial, dangers which I believe to be wholly imaginary.

If our cemeteries are to be replaced by furnaces with tall chimneys pouring their products of combustion into the air, it is self-evident that the cities will gain nothing by the reintroduction of a mode of disposal of the dead which, whatever may be its merits, is not in accordance with the teachings of modern science.

To sum up: it appears, that, as compared with cremation, inhumation is cheaper, simpler, and quicker. It is *productive* and not *destructive*, it is indirectly a cause of freshening the air instead of fouling it, and provides a lovely spot for the enjoyment of the living.

CHAPTER XIII

THE STORY OF BREMONTIER, AND THE RECLAMATION OF THE SAND-WASTES OF GASCONY[1]

In the short address which I have the honour to give to you this evening, I purpose to bring to your notice the chief facts of a great sanitary work which has been accomplished by our friends and neighbours the French.

If you will take the map of France and look at that portion of the coast which skirts the Bay of Biscay, you will notice that two great rivers flow into the sea along this coast. One, the most northerly, is the Gironde, a stream which has upon its banks the great commercial city of Bordeaux; the other river is the Adour, the mouth of which is 150 miles south of the mouth of the Gironde.

Between the mouths of these two rivers the shore of the Bay of Biscay is formed absolutely and entirely of sand, and for a considerable distance inland from the coast the soil of France is composed of sand. It is to this great sandy district, covering nearly two millions and a half of acres, and known in France as the ' Landes ' or Moorlands, that I wish to direct your attention.

These moorlands have been the despair of agriculturists for centuries, and have been universally regarded

[1] This chapter formed the subject of an address to the working classes delivered at York in 1886 on the occasion of the Congress held by the Sanitary Institute. Its original form has not been altered.

as among the dreariest and most unwholesome districts in Europe. Sand has not the reputation of being a very profitable soil to the agriculturist, and in addition to the natural poverty of the soil the farmer in this region has had to contend with the impossibility of efficient drainage. The Landes formerly produced nothing except a scant herbage sufficient to support a few miserable sheep, tended by shepherds as ill-favoured as their flocks, who generally suffered from one or other of the many diseases prevalent in the country; for disease was about the only crop which the Landes formerly brought forth abundantly.

Indeed, you will find that plains which are unproductive are generally unhealthy. The Campagna round Rome is a very hotbed of malarious and other diseases, and the sandy plains of Holland, and our own Lincolnshire, enjoyed a similar evil repute, before efficient drainage was brought about by skilful engineers, and the cultivation of the soil became possible. Husbandry and disease are sworn foes, and the pursuit of agriculture is generally the pursuit of health, and a healthy man is generally contented. Here is an argument for 'small holdings,' for 'three acres and a cow,' and for 'allotments,' which I freely give to those who find pleasure in political contention.

The drainage of the Landes presented special difficulties, and difficulties which no engineering skill and no expenditure of money in the direction of bricks, mortar and machinery seemed likely to overcome, and for the following reasons:—

The reputation of the Bay of Biscay is familiar to every Englishman. It is there, if anywhere, that the force of wind asserts itself, and the winds are generally westerly in direction, and blow with fearful violence from the sea over the land. The shore of that part of the bay with which I am dealing is composed, as I have said, of unmitigated sand. The effect of the wind upon sand is familiar to all of us, for the sand is borne before the wind and travels considerable distances.

Now, in the Bay of Biscay the rise and fall of the tide is great, so that the sand washed up by the sea is left high and dry to the extent of many feet at low water.

Again, in the latitude of the Bay of Biscay the sun is far more powerful than here, so that in the interval between the times of high water the sand is greatly heated by the sun, and is so thoroughly dried that the particles no longer tend to stick together—glued by natural moisture—but are easily driven before the furious blast which comes roaring from the sea. When the wind is not very strong it blows the sand into heaps along the shore. These heaps or hills may reach an elevation of from 60 to 300 feet, with an inclination of about 30 degrees towards the sea. These heaps of sand are called 'dunes,' a word having the same origin probably as the English 'down,' and formerly the whole foreshore of the Bay of Biscay, between the Gironde and the Adour, presented an undulating appearance, as though a portion of the swelling, rolling sea had been turned to sand and become stationary. If these sandhills had been really stationary they would have formed a natural rampart against wind and waves, and it might have been possible to drain and cultivate the land behind them. But this was not the case. The scanty herbage of grass and reed which grew upon the dunes was not enough to fix them. It only required a gale of moderate force to completely alter the face of the country: hills became flat, valleys were filled up, the

THE STORY OF BRÉMONTIER

lakes which formed behind the dunes became dry land, the water which the lakes contained was forced in some new direction, and what happened to the lakes also happened to the watercourses, with the result that the whole country was waterlogged, and fields and gardens which had been painfully and industriously cultivated were submerged by the drifting sand. It is even stated that villages disappeared completely in this way, and that an enterprising agriculturist in digging his estate was surprised at finding just beneath the surface the brazen weather-cock on the steeple of a long-forgotten parish church! It is a great labour, even at the present day, to keep the mouths of the Gironde and the Adour free from drifting sand, and it is certain that a century or so ago the course of the Adour was completely changed, owing to the channel getting dammed by sand blown into it. If an accident such as this could happen to a mighty stream like the Adour, one may judge of the great uncertainty which attended the course of smaller streams, and the absolute impossibility of draining the land.

A few feet below the average level of the surface of the district there is an impermeable stratum, locally known as *alios*, which keeps the water from flowing away, and beneath the impermeable stratum was more sand sodden with undrinkable water.

The result of this condition of things naturally was that the district of the Landes during the wet season was a swamp, and during the dry season a pestilential morass. The district was uncultivated, and produced nothing but scanty herbage, which served as pasture for a few wretched sheep, tended by shepherds doomed to spend their lives upon stilts, for the country was such that it was impossible to walk far in any one direction without sinking to the waist or shoulders. The country produced no corn and the population was the scantiest in proportion to acreage of any district in France. The population was kept down also by disease. Fevers of all kinds—and especially those of a malarious type—were exceedingly common. And in addition, there was a disease peculiar to this and a few other districts in Europe, known as Pellagra; a terrible disease, which disfigured and slowly killed; the patient dying with the aspect of a mummy and the mind of an imbecile.

The Landes had remained for centuries as a blemish on the fair face of France, and all attempts to reclaim and cultivate them had signally failed. The Emperor Charlemagne, it is said, employed his troops in the intervals of his Spanish campaigns in an attempt to reclaim the Landes, but the forces of nature laughed at the puny opposition of the greatest magnate of the world, and at once resumed their sway as soon as the imperial soldiers had ceased to dig ditch and throw up bank.

I have no fear of being contradicted when I say that it is of no use to attempt to fight with Nature. We may oppose her for a time, but only for a time. In the end she asserts her sway, and man sees too late how his labour has been in vain.

Dwellers in these islands do not need to be reminded of the awful and irresistible power of wind and wave, against which the mere deadweight of cyclopean breakwaters, constructed at gigantic cost and maintained by constant periodic expenditure, is at times laughably impotent. I need not say that the wind and waves of the Bay of Biscay are the roughest and rudest in the world, and that if the maintenance of dead breakwaters is an endless and almost hopeless task on our coasts, on the stretch of coast which I am considering their construction and maintenance would be alike impossible. Thus it was that until the latter end of the last century the condition of the Landes, a tract of two millions and a half of acres, seemed hopeless, and they seemed doomed to be open to the fury of sand storms for ever and to remain a pestilential, unprofitable, undrained swamp to all eternity.

But, happily for France, and especially for the dwellers between the Gironde and the Adour, there was born in 1738 Nicolas Thomas Bremontier. It is said that the world knows nothing of its greatest men. Certain it is that Bremontier was one of the greatest benefactors to humanity that the world has ever known, but I regret to say that I can tell you very little about his life. This would have been less but for the kindness of Mr. Jenkins, the late Secretary of the Royal Agricultural Society, who obtained from Paris the following extract from the memoirs of the Agricultural Society of the Seine. This short biographical notice of Bremontier is from the thirteenth volume of the Transactions of the Society (for the year 1810), and was most courteously extracted by M. Laverrière, the librarian:—

'Nicolas Thomas Bremontier was born at Quavilly, near Rouen, July 30, 1738, and soon manifested great aptitude for the exact sciences. He was very young when he entered the school of the Ponts et Chaussées, and at 18 he went to the College of the Marine Artillery at Toulon, to teach applied mathematics. This school, established by M. Choiseul, was broken up a few years later, and Bremontier went as Engineer of Roads and Bridges, first to Perigueux, and then to Bordeaux. Here he was actively engaged in his profession, and published papers on the drainage of marshes in the neighbourhood of Bordeaux, on the cleansing of the Bordeaux harbour, and on the methods of restraining rivers and torrents to their proper beds. His energy was inexhaustible, and in his leisure he taught himself the principles of music, and became in this direction most efficient. Promoted to be Inspector of Roads and Bridges, he went to Brittany to make a canal to join the Rance to the Villaine. Thence he was sent to Normandy to make a canal from the Orne, by Caen, to the sea. At Caen, he reconciled the differences which had arisen between the provincial magnates and the officers of the Ponts et Chaussées, and his judgment and conciliatory spirit had proved useful in a similar way in Bordeaux. When, therefore, the post of Engineer in Chief for Guienne became vacant, he was appointed in obedience to the wishes of the locality.

'Bremontier joyfully accepted this post, not merely because it was at once a professional and social promotion, but mainly because while formerly living at Bordeaux he had been a witness of great troubles for which he believed he had found a remedy; at least his early experiments (conducted at his own cost) gave him a strong cause for hope. Bremontier felt that here was to be the theatre of his greatest and most useful labours. Possibly, we might believe, he thought to earn an imperishable fame, for self-interest, we are prone to think, is the main spring of good works. Bremontier had less need of such a spur than most men.

'He had visited the sandhills of Gascony during his first sojourn at Bordeaux, and bewailed the misery caused by those moving mountains thrown up by the sea and driven by the west wind, which had already smothered a vast tract of cultivated land, as well as rural inhabitants and villagers, and threatened to cover the more fertile districts and advance even to Bordeaux itself.

'The idea of arresting this devastating power took possession of Bremontier, and the hope of success occupied entirely his brain and hands. He studied the nature and the movements of the sand, he measured their extent, and noted the ravages past and to come.

'He recognised their vegetative power, and from the year 1787 he knew that a great number of plants, and especially resinous trees, could find nourishment in them. He made experiments at his own cost to get some definite facts. He perfected his method of procedure, and at last, certain of success, and feeling that such an enterprise was beyond the power of a single man, he sought the help and succour of the Government. His assertions were not credited, and his project shared the fate of many other creations of genius which are repelled at their birth by the ignorant, until the results become so numerous and evident as no longer to be neglected.

'After Bremontier's first attempt the solid basis of procedure was found, and the extension of his work alone was necessary. What proportion was there between the few acres planted at the cost and by the care of one man, and that vast stretch of country extending from the Gironde to the Adour, nearly 180 miles, and averaging three or four in breadth, all exposed to the action of the destructive sand and in part covered by it?

'During his second residence at Bordeaux he renewed his application to Government, this time still more certain of the success of his project.

'In a short notice like this it is not possible to enter into all his trials and difficulties, nor to dilate upon the dangers to which he and his works were alike exposed during the period of anarchy to which France at that time was so long a prey.

'The genius who then controlled the destinies of the Empire appreciated the value of Bremontier's projects, and in the year 1801 he allotted 50,000 francs for the continuance of the work, and a similar sum has been allotted to it in each succeeding year.

'Bremontier now began to enjoy the fruits of his labour, and in 1808, 3,700 hectares of land (about 9,000 acres) had already been sown. Honoured by the esteem of the department of Ponts et Chaussées, he had been promoted to the rank of Inspector-General, and he was chosen by his fellow-citizens of Bordeaux to be one of a deputation to wait upon the Emperor at Bayonne. He had then the happiness to submit to this great ruler his future projects and his past success, and felt assured that from that time the great work to which he had devoted himself would not be abandoned, and that its future success was assured. The fixation of the whole of the Dunes is now ranked among the great public benefits to which Napoleon with the instinct of a genius gave his support. Bremontier in his dying hours was doubtless consoled by this pleasing prospect, and he breathed his last surrounded by his friends, and with the calmness and resignation of a true philosopher.'

Bremontier recognised the fact that the only way to grapple with the forces of Nature is, not to fight blindly with them, but to try to make use of them. Nature is always working for our benefit, and although it seems as though at times, in a fit of anger as it were (the real object of which we may fail to comprehend), she destroys much of her own work, still, in the long run, those who endeavour to turn the forces of Nature to account will find the balance enormously in their favour.

It is well known, and has been long recognised, that the best protection for a bank or rampart against the fury of the elements is to plant it. A loose heap of earth is liable (no matter how huge it may be) to be washed and blown away in times of tempest. If, however, the bank be planted, the roots of the trees and plants hold the elements of the soil together, and the spreading branches and leaves form at the same time a protection from the fury of wind and water. It is true that even planted hills and banks may suffer severely in times of exceptional storm, but the storm once past, the silent forces of Nature commence at once the work of reconstruction; the damaged roots send forth fresh rootlets, the damaged branches soon push again with buds of promise, and possibly before the advent of the next exceptional gale, the storm rampart is stronger than before. These silent forces of Nature are truly beneficent; they merely ask for fair play, they work for us without wage; and one great principle of success in all work in this world, be it legislative, be it sanitary, or be it of any other kind, is to go with them, not to fight against them, to learn if we can what is Nature's inexorable law, and lay to our hearts the fact that Nature brooks neither stubbornness nor disobedience.

Bremontier recognised the fact that the only way to fix the drifting sand-dunes was to plant them: but how, and with what? These were the questions he had to solve. Sand is not regarded as a promising soil by agriculturists in general, and the sea-sand along the shore the least promising of all. And yet sand must contain in its interstices a good deal of organic matter left by the seaward-tending rivers, and the fact that the sandy estuaries of rivers are very liable to breed malaria may be taken as evidence that organic matter must exist in quantity and in fine division among the minute particles of sand. This spring (1886) I was astonished at finding, close to Biarritz and within a few yards of the sea, a very flourishing crop of peas which had been sown in the sand without, apparently, the admixture of any manurial body. They were protected from the sea-winds by hurdling made of gorse, and enjoyed an ample exposure to the sun, and thus bid fair to yield a good return in due time. The pea is a plant that sends its roots very deeply, and the roots doubtless found moisture and nourishment at a great depth below the surface. For fixing dunes, however, something more permanent than peas is necessary, and Bremontier resolved to try the *Pinus maritima*, a tree which was known to flourish in sandy soils near the coast. The *Pinus maritima* is a species of Pinaster, and in habit and size it very much resembles the common 'Scotch Fir,' with which we are all familiar. Bremontier made his first sowings of the seeds of *Pinus maritima* in the year 1789, and I will state shortly his perfected manner of procedure by which he overcame the obvious difficulties of his task.

I wonder what the dull-minded and prejudiced peasant thought of this enthusiast who went forth to do battle with the mighty ocean, and still mightier wind, armed only with a few handfuls of pine seeds, such as might be driven far away by the first strong gust that blew. I wonder also if only the ignorant laughed at him, and if he escaped the jeers and sneers of those who had enjoyed the advantages of a better education. Probably not, and equally probably he cared little for the opinions of the prejudiced. The pine seeds were sown mixed with seeds of the common broom, and the sowings were made in a direction at right angles to the prevailing wind. A screen of hurdles made of gorse or of planks deeply driven into the sand was placed on the windward side of the seed-ground, and the seed-ground itself was thatched with pine branches and other suitable material. At the end of the first year, the broom would be nine or ten inches high and the pine saplings only two or three inches, and thus the tender little saplings were nursed and protected by the plants of broom. In half-a-dozen years or so the brooms had reached their full growth, but the pines continued to grow, and, in course of time, overtopped the brooms, and smothered their nurses. Being judiciously thinned and pruned by the foresters, the pines grew into fine trees able to resist the fury of the elements, sending their long tap-roots and laterals in all directions through the dunes, and causing them to become year by year a stronger and stronger protection to the inland wastes instead of a dangerous menace. Before the dawn of the present century Bremontier had proved the success of his practice. In the year 1801 the matter was taken in hand by the French Government, and in 1810 it was ordained that so much of the sand-dunes as belonged to the State should be planted after the manner of Bremontier, while the private property of those who were unwilling or unable to plant should be taken in hand by the State, all revenue arising from such land being confiscated until the cost entailed by the work had been recouped.

In 1817, a yearly sum of less than 4,000*l*. was voted for the reclamation of the dunes and wastes of Gascony; with the result that, in the department of Landes, 98,000 acres of forest have been planted; and whereas in 1834 there were about 900,000 acres of uncultivable land in the department of Landes alone, there are now only 340,000 acres, showing that in the past half century reclamation has proceeded at the rate of 12,000 acres a year. These figures apply only to the department of the Landes, and leave out of consideration the department of Gironde, in which, however, nearly half these waste moorlands are situated. This reclamation has been made possible by the fixation of the dunes, which has rendered systematic drainage operations practicable. Canals and drains have been cut in every direction, and, thanks to the pine forests, there is now no longer any risk of their being choked up with sand.

The *Pinus maritima* has proved a very profitable tree. Within twenty or twenty-five years of sowing, it begins to yield a return. The timber is of very moderate quality, but is largely used for packing-cases, for shores in the dockyards of Bordeaux, for railway sleepers, and for firewood. I may remark in passing, that the great scarcity of coal in France compels the French to look to their forests for fuel, and there is probably no nation more clever and more thrifty in the management of trees.

The pine trees are chiefly valuable for their yield of turpentine and resin, which in that comparatively warm climate is very abundant. The resin is obtained by removing a strip of bark from the tree and allowing the exuding sap to trickle into a small earthen vessel shaped like a flower-pot. The trees begin to yield resin when they are about twenty years old, and the resin is worth about 5*l*. a hogshead in its raw crude condition. As far as I am able to judge, it requires about 250 trees on an acre of ground to give a hogshead of resin. It requires comparatively little labour to collect the resin, so that the profit per acre from the resin harvest is considerable. It is said that the draining away of the resin does not seriously affect the value of the timber. Besides resin and timber, the manufacture of charcoal is largely carried on—charcoal, as you are aware, being in great demand in France for a variety of purposes.

Thus it appears that the waste moorlands on the shores of the Bay of Biscay have become of great commercial value. Journeying from Bordeaux to Bayonne, the railway passes through one long monotonous pine forest. When I state that the journey takes between four and five hours, you will be able to judge of the vast tract of country which, once the abomination of desolations, is now covered with millions of the resin-yielding *Pinus maritima*. The cultivation of the pine improves the soil, which is gradually enriched and altered in quality by the dead leaves and other vegetable débris which fall upon it. In some places clearings have been made in the forest and vineyards planted, and I need not remind you that the most valuable vineyards in the world are on the southern bank of the Gironde, on the very fringe of the pine woods which I have been describing.

The rise in agricultural value of this tract of country, great as it is, is a small matter. The great gain after all has been the rendering wholesome of a pestilential swamp and the removal of a plague spot from the face of Nature. The shepherds of the Landes, except in very few places, have now no longer any need to walk about on stilts, and malaria and pellagra from being common have become rarities, and will soon be extinct. Life in this district no longer languishes and ends prematurely, but the inhabitants enjoy a vigorous health, and that happiness and contentment which vigorous health alone can give.

Population has increased very rapidly since the

beginning of the century, and industries of various kinds are able to be carried on. Round the basin of Arcachon is a very large population supported mainly by the oyster fisheries, and the town of Arcachon, which has grown up in the pine forest, is one of the well-known health resorts in Europe, where land in the best situations is worth about 1,000*l.* an acre. Well may the dwellers in Arcachon raise a statue to Bremontier, whose far-seeing and thrifty policy has brought them health, happiness, and riches in place of disease, misery, and poverty.

I have now given you the simple details of the manner in which Bremontier's small beginning has made great end; how his pine plantations, made at first with no little labour and sorrow, began along the coast, and with the lapse of a century have reclaimed a province.

You will be asking, perhaps, why I have chosen this subject for my short address to the inhabitants of York, and having listened to my tale, you will be asking for the moral.

I chose this subject for my address for several reasons. The chief reason probably is to be found in the fact that I spent part of the early spring of this year in the district which I have been describing, and what I saw there made, as it could not fail to do, a very deep impression upon me.

My next reason was that it is an aspect of sanitation which is not often dealt with at meetings like this, and I was glad of the opportunity of taking you away from pipes, traps, sinks, and those expensive roads to health which we have to consider in cities, to contemplate the sanitary effect of good husbandry in the open-air; and to show you on a large scale what I believe to be universally true, viz. that the cultivator of the soil must always be the right-hand man of the sanitarian.

It has been refreshing for us to contemplate a sanitary work which has proved a financial success. Sanitation always gives us the best of all dividends—health. It is said to be a short-sighted policy, especially in cities, to look for a money return on the capital expended on works for improving the public health. The thrifty French, however, have given to the world a valuable example of a comparatively small expenditure yielding *in the course of time* a magnificent return of both health and material prosperity.

Do not run away with the idea that the *Pinus maritima* is a cure for all waste lands and unwholesome districts, because it happens to be especially suited for the soil and climate of the eastern shores of the Bay of Biscay. In the warm climate of the south it yields abundance of resin and turpentine, grows quickly, and furnishes a large quantity of timber. In more northern climates it will grow, but does not flourish; and although there is at least one fine specimen in Kew Gardens, it is not, from all I have heard, a tree suited to this climate.

My story seems to show that in the reclamation of waste lands we must not be in a hurry. Nature is sure, but from our point of view, slow. Bremontier, and those who worked with him, began in a small way. We may be sure that experience had to be bought at more or less expense; and it was not until the success of his methods had been proved that the French Government seriously took the matter in hand. Bremontier was a true patriot. He worked solely for the good of his country and for posterity. He had no idea of immediate profit, either for himself or his contemporaries. He drew his modest salary as Inspector-General of roads and bridges, but looked to no further profit. He lived barely long enough to see the resin flow from his first plantings. He pointed out, as it were, the way to the promised land, but for himself he only saw the promised land 'in his mind's eye.' It is good for us to bear this fact in mind, for many reformers of the present day seem, in questions of land management, to look only for immediate results, and to be actuated by the not very noble sentiment of 'bother posterity, what has posterity done for me?'

There has been a good deal of talk of late about the reclamation of waste lands in this country, and the opinion of some seems to be that worthless soil presents a glorious opportunity of wasting money. These are questions concerning which I cannot speak to you as an expert, but it seems certain that the problem of reclamation must differ with the circumstances of soil and situation, and that it is far more easy to do the wrong thing than the right. The first thing necessary is to find a Bremontier to show the way. We shall want a Bremontier to show us the way out of the pestilential quagmire which we Londoners are making by dint of large expenditure in the estuary of the Thames. We want a genius and enthusiast who will do for the bogs of Ireland what this great Frenchman did for the Landes of Gascony.[1]

[1] For many of the facts embodied in this address I am indebted to Dr. John Croumbie Brown's 'Pine Plantations on the Sand-wastes of France.' Edinburgh (Oliver and Boyd), 1878.

APPENDIX

The following paragraphs are added in illustration of certain points which have been discussed in the body of the work:—

OVERCROWDING AND AIR-BORNE CONTAGIA.

In the second chapter it has been pointed out that overcrowding in cities is especially dangerous in relation to air-borne contagia. This is very well shown by the effect of the recent influenza epidemics upon the death-rates of the 'Central districts' of the Registrar-General as compared with the death-rates for London as a whole.

We have had recently in London three epidemics of influenza: the first occurred in the first quarter of 1890, the second in the second quarter of 1891, and the third in the first quarter of 1892. The following table gives the true death-rates (per 1,000 per annum) for these three quarter years in London as a whole, and in each of the Central districts of the Registrar-General. Comment is unnecessary.

True London Death-rates during three Epidemics of Influenza.

	First quarter, 1890	Second quarter, 1891	First quarter, 1892	Average of three epidemics
London as a whole .	23·2	23·5	27·8	24·8
St. Giles'	33·1	31·0	35·1	33·1
St. Martin's . . .	26·7	27·5	27·7	27·3
Strand	31·7	36·3	44·0	37·3
Holborn	38·8	37·1	36·6	37·5
Clerkenwell . . .	28·1	36·7	35·3	30·0
St. Luke's	30·9	31·8	35·5	32·7
City	36·8	30·8	33·0	33·5

THE NIETLEBEN OUTBREAK OF CHOLERA.

(Showing the occasional dangers of Sewage Farms.)

It will be within the recollection of most readers that in January 1893 a sharp outbreak of cholera (affecting 115 persons, and causing 40 deaths) occurred in a lunatic asylum at Nietleben, on the Saale, a village not far from Halle.

In the *British Medical Journal* for February 4, 1893 (p. 250), is the following note from the correspondent of the *Journal* at Berlin: 'The Nietleben epidemic seems a further proof that contaminated water is the cause of cholera. The water arrangements of the asylum are, on a small scale, similar to those of Hamburg. The asylum drains are carried to a sewage farm, the overflow water from this latter to the River Saale, and close below the mouth of the pipe lies the tube which carries Saale water to Nietleben. It is true that the water, before reaching the asylum, passes over a sand filter, which is an improvement on the state of things in Hamburg; but during the late and long-continued frost not only were the filter-beds quite frozen, but the sewage farm, too, was inactive. Thus, during the cold weather, the drain-water from the establishment went into the Saale unpurified, and returned to it enormously diluted, it is true, by river water, but absolutely unfiltered. No one will be surprised to hear that amongst the unfortunate drinkers of this diluted sewage epidemic diarrhœa broke out. From where and how the first bacilli of cholera asiatica reached the asylum has not yet been cleared up, though there has been much talk of a male nurse from Hamburg.'

In the same number of the *Journal* it is stated that the bacilli of cholera had been found in the water supplied to the asylum. It is further stated that several cases of cholera had occurred at Trotha, a village on the right bank of the Saale, and two miles below the asylum.

MARKET-GARDENING.

Almost the only branch of British agriculture which is still in a fairly prosperous condition is market-gardening, and few who have not lived in the suburbs of London can form any idea of the vast importance of this industry from the economic and sanitary points of view.

William Cobbett, in his 'English Gardener,' says (speaking of the London market-gardeners): 'These gardeners excel all the world in everything that they undertake to cultivate; they beat all the gentlemen's gardeners in the kingdom; nothing ever fails that depends upon their skill; and I should be ungrateful, indeed, if I did not acknowledge that I have learned more from them than from all the books I have read in my life, and from all that I ever saw practised in gentlemen's gardens.'

This eulogium still holds good; and, indeed, the quantity of food that is raised on the market-gardens round London is simply astounding.

The comparative prosperity of the London market-gardeners is due to two facts: the proximity of an insatiable market, and the possibility of getting an almost unlimited amount of dung from the London stables. This dung is very cheap, because the Londoner is obliged to get rid of it at any price, and in many instances he even pays in order to have it removed.

The quantity of dung which is put upon the London market-gardens is prodigious, and although the amount varies with the crop to be raised, the average is stated, on the authority of two well-known gardeners, to range between sixty and eighty tons per annum to the acre. The amount used for raising onions is very large, and it is larger still on the grounds devoted to growing gherkins and ridge-cucumbers. I am given to understand that it is possible to put as much as 200 tons of stable manure on an acre of ground, although it would be difficult to get any return from land which is of such a hungry nature as to take such a large quantity. These figures are of interest as showing how enormous is the power of cultivated land to appropriate organic manure, and how groundless are any fears as to the land becoming foul provided it be laboriously tilled.

The rent paid for market-garden ground very commonly amounts to ten pounds per acre, and if the ground be planted with orchard trees, as much as fourteen pounds per annum may be obtained.

The capital involved in market-gardening is very great, and the plant required—store-houses, tools, carts, horses, baskets, glass-houses and lights—is far more complex and costly than that required for farming.

The labour necessary for market-gardening is very large indeed, and many of the gardeners in the district which I have in my mind pay as much as 300$l.$ a week for wages in the summer, and about half that amount during the slack season.

The traffic on the main roads which lead from the gardening districts to London is astonishingly great, and the stream of carts wending towards market with their loads of produce, or returning from market with their loads of dung, never ceases during the whole of the twenty-four hours. When one considers the high rent, the large sums paid for wages, and the enormous amount of manure used (which, I believe, costs about half-a-crown per ton delivered on the land), it is evident that the yield must be infinitely greater than that obtained by the farmer; were it otherwise, no profit could be made. No better instance than this could be given of the magnificent results obtained by the scientific use of organic refuse.

The London stables are, as a rule, the perfection of cleanliness, because the dung, which is capable of being turned to very profitable use, is eagerly sought after and removed. The London market-gardens are the most productive in the world, because the stables are clean. This is hygiene worthy of imitation.

It is to be hoped that the Yahoos may never succeed in inducing the Houyhnhnms to adopt their Augean system.

INDEX

ACRE crop, value of an, 84
Agricultural degradation, 190
Air, expired, 142
— fresh, 146
— influence of plants on, 147
— spread of disease through, 148
— London, 155
Andover, sanitation in, 221
Antiseptics, 69
Areas of London, 16
Ashes, use of, 207

BARRACKS, construction of, 133
Berlin, sewerage of, 98
Boussingault on nitrates, 195
Bremontier, story of, 298
Burial, 285
— cost of scientific, 295
— v. cremation, 39, 293

CEMETERIES, 287
Cesspools, 173
Chicago, 28
China, population of, 179
Chinese sanitation, 107
Cholera, 63, 67
— bacillus, 93
— — destruction of, 94
— Nietleben outbreak of, 316
Churches, ventilation of, 144
'City,' population of the, 36
Concentration of population, America, 10

Concentration of population, England, 11
— — — London, 181
— — — cause of, 23
— — — evils of, 13, 190, 315
— — — and influenza, 22
Cremation, 290
Curtilage, proper, 42

DEATH-RATE, Dorsetshire, 77
— Edinburgh, 25
— London district, 77
— London proper, 17
— through London fog, 18
— New York, 19
— factory, 15
— town and country, 14
Disinfection, practical details of, 233
Drains, dangers of house, 123
Drinking water, Frankland on, 245
— — meaning of nitrates in, 250

EARTH, as a filter, 111, 242
— a natural steriliser, 115
— experiments on living, 112
— tilling of the, 86
— to earth burial, 286
Earth-closet, 198, 214
— earth as manure, 200
— practical details of, 204
Earthworms, use of, 87
Excrement, removal of, 193

FER

Fermentation, 86
Filtration through earth, 90, 242
'Flats,' evils of, 25, 31
Fog and death-rate, 156
— — infection, 151
— increase of, 33
— London, 18

Garden, practical details in manuring, 229
Gilbert and Lawes on manuring, 57

Health in Chicago and London, 27
Hotel, a model, 139
Hotels, construction of, 133
— evils of, 31, 134
— ventilation of, 137
House and garden, 47
— construction of the, 119, 130
'Humification,' 203
Humus, 91, 202
Hygienic units, 176

Infection, and concentration of population, 21, 315
Influenza, and overcrowding, 40

London air, 155
— district, death-rate of, 77
— — lung diseases in, 80
— — streams and watershed, 186
— — water-supply, 190

Manures, chemical and natural, 60, 201
— value of excrement as a, 102, 228
— waste of, 74
— — — sewage, 103
— — — urine, 103
Manuring of land, 55
Market-gardening, 317
Micro-organisms in earth, 51

SEW

Model dwellings, 45
Modern mansions, 34

Nature, an adjuvant in reclamation, 306
Nitrates, formation of, 62, 195
— in water, meaning of, 250
Nitrification, 51, 88, 91, 286
— of pathogenic organisms, 92
Nitrifying organisms, factors influencing growth of, 54, 88, 91
— — Frankland's, 89
Northumberland Avenue, 35

Open-air spaces, 38
Overcrowding, 40, 75
— legislation against, 41
Oxydation, 63

Pail system, 193
Percolation of sewage, 72
Population, acreage for, 191
— of the 'City,' 36
Putrefaction and oxydation, 63

Refuse, kitchen, 205
— trade, 209, 283
— utilisation of, 121
Richardson on w.c., 210
River pollution, 159, 235
— — Acts against, 160
— — Ruskin on, 157

Sanitary economy, 84
— legislation, 83
Sanitation in Andover, 221
— a branch of agriculture, 96
— by the individual, 109
— Chinese, 107
— local board regulations for, 264
— model scheme for, 225, 279
— suburban, 260
Schools, construction of, 132
Sewage, chemicals in town, 231
— disposal of, 105
— purification of, 97

INDEX

SEW

Sewage, without excrement, 100
— evils of water-carried, 81, 211
— schemes, cost of, 73, 274
— — false economy of, 74, 275
— — reform of, 279
Slops, use of, 105, 256, 258
Small-pox infection, 149
Soil, starvation of, 109
— value of top-, 90
Strand district, statistics of population of, 78

THAMES, fouling of, 71, 261
Typhoid fever, 161
— epidemic, Caterham, 162
— — Lausen, 163
— epidemics, 99
Typhus, 63

WEL

URINE, manurial value of, 116

VEGETABLE mould, 87
Ventilation, 143, 146, 153

WATER, meaning of nitrates in, 250
— pollution of, 163
— purification of, 171
— -supply, 175, 182, 237
— — London, 67, 71, 190
Water-closet, 99, 63, 121, 131
— Richardson on, 210
Wells, contamination of London, 95
— danger of deep, 168
— pollution of, 71, 167
— surface, 174, 242

PRINTED BY
SPOTTISWOODE AND CO., NEW-STREET SQUARE
LONDON

Y

A CATALOGUE OF WORKS
IN
GENERAL LITERATURE
PUBLISHED BY
MESSRS. LONGMANS, GREEN, & CO.,
89 PATERNOSTER ROW, LONDON, E.C.,

15 EAST 16TH STREET, NEW YORK.

MESSRS. LONGMANS, GREEN, & CO.
Issue the undermentioned Lists of their Publications, which may be had post free on application:—

1. MONTHLY LIST OF NEW WORKS AND NEW EDITIONS.
2. QUARTERLY LIST OF ANNOUNCEMENTS AND NEW WORKS.
3. NOTES ON BOOKS; BEING AN ANALYSIS OF THE WORKS PUBLISHED DURING EACH QUARTER.
4. CATALOGUE OF SCIENTIFIC WORKS.
5. CATALOGUE OF MEDICAL AND SURGICAL WORKS.
6. CATALOGUE OF SCHOOL BOOKS AND EDUCATIONAL WORKS.
7. CATALOGUE OF BOOKS FOR ELEMENTARY SCHOOLS AND PUPIL TEACHERS.
8. CATALOGUE OF THEOLOGICAL WORKS BY DIVINES AND MEMBERS OF THE CHURCH OF ENGLAND.
9. CATALOGUE OF WORKS IN GENERAL LITERATURE.

ABBEY (Rev. C. J.) and OVERTON (Rev. J. H.).—THE ENGLISH CHURCH IN THE EIGHTEENTH CENTURY. Cr. 8vo. 7s. 6d.

ABBOTT (Evelyn).—A HISTORY OF GREECE. In Two Parts.
Part I.—From the Earliest Times to the Ionian Revolt. Cr. 8vo. 10s. 6d.
Part II.—500-445 B.C. 10s. 6d.

—— A SKELETON OUTLINE OF GREEK HISTORY. Chronologically Arranged. Crown 8vo. 2s. 6d.

—— HELLENICA. A Collection of Essays on Greek Poetry, Philosophy, History, and Religion. Edited by EVELYN ABBOTT. 8vo. 16s.

ACLAND (A. H. Dyke) and RANSOME (Cyril).—A HANDBOOK IN OUTLINE OF THE POLITICAL HISTORY OF ENGLAND TO 1890. Chronologically Arranged. Crown 8vo. 6s.

ACTON (Eliza).—MODERN COOKERY. With 150 Woodcuts. Fcp. 8vo. 4s. 6d.

A. K. H. B.—THE ESSAYS AND CONTRIBUTIONS OF. Crown 8vo. 3s. 6d. each.

Autumn Holidays of a Country Parson.
Changed Aspects of Unchanged Truths.
Commonplace Philosopher.
Counsel and Comfort from a City Pulpit.
Critical Essays of a Country Parson.
East Coast Days and Memories.
Graver Thoughts of a Country Parson. Three Series.
Landscapes, Churches, and Moralities.
Leisure Hours in Town.
Lessons of Middle Age.
Our Little Life. Two Series.
Our Homely Comedy and Tragedy.
Present Day Thoughts.
Recreations of a Country Parson. Three Series. Also 1st Series. 6d.
Seaside Musings.
Sunday Afternoons in the Parish Church of a Scottish University City.

—— 'To Meet the Day' through the Christian Year; being a Text of Scripture, with an Original Meditation and a Short Selection in Verse for Every Day. Crown 8vo. 4s. 6d.

—— TWENTY-FIVE YEARS OF ST. ANDREWS. 1865-1890. 2 vols. 8vo. Vol. I. 12s. Vol. II. 15s.

AMOS (Sheldon).—A PRIMER OF THE ENGLISH CONSTITUTION AND GOVERNMENT. Crown 8vo. 6s.

ANNUAL REGISTER (The). A Review of Public Events at Home and Abroad, for the year 1891. 8vo. 18s.

*** Volumes of the 'Annual Register' for the years 1863-1890 can still be had.

ANSTEY (F.).—THE BLACK POODLE, and other Stories. Crown 8vo. 2s. boards.; 2s. 6d. cloth.

—— VOCES POPULI. Reprinted from *Punch*. With Illustrations by J. BERNARD PARTRIDGE. First Series, Fcp. 4to. 5s. Second Series. Fcp. 4to. 6s.

—— THE TRAVELLING COMPANIONS. Reprinted from *Punch*. With Illustrations by J. BARNARD PARTRIDGE. Post 4to. 5s.

ARISTOTLE—The Works of.

—— THE POLITICS, G. Bekker's Greek Text of Books I. III. IV. (VII.), with an English Translation by W. E. BOLLAND, and short Introductory Essays by ANDREW LANG. Crown 8vo. 7s. 6d.

—— THE POLITICS, Introductory Essays. By ANDREW LANG. (From Bolland and Lang's 'Politics'.) Crown 8vo. 2s. 6d.

—— THE ETHICS, Greek Text, illustrated with Essays and Notes. By Sir ALEXANDER GRANT, Bart. 2 vols. 8vo. 32s.

—— THE NICOMACHEAN ETHICS, newly translated into English. By ROBERT WILLIAMS. Crown 8vo. 7s. 6d.

ARMSTRONG (Ed.).—ELISABETH FARNESE: the Termagant of Spain. 8vo. 16s.

ARMSTRONG (G. F. Savage-).—POEMS: Lyrical and Dramatic. Fcp. 8vo. 6s.

BY THE SAME AUTHOR. Fcp. 8vo.

King Saul. 5s.
King David. 6s.
King Solomon. 6s.
Ugone; a Tragedy. 6s.
A Garland from Greece. Poems. 7s. 6d.
Stories of Wicklow. Poems. 7s. 6d.
Mephistopheles in Broadcloth; a Satire. 4s.
One in the Infinite; a Poem. Crown 8vo. 7s. 6d.
The Life and Letters of Edmond J. Armstrong. 7s. 6d.

ARMSTRONG (E. J.).—POETICAL WORKS. Fcp. 8vo. 5s.

—— ESSAYS AND SKETCHES. Fcp. 8vo. 5s.

ARNOLD (Sir Edwin).—THE LIGHT OF THE WORLD, or the Great Consummation. A Poem. Crown 8vo. 7s. 6d. net.

Presentation Edition. With Illustrations by W. HOLMAN HUNT. 4to. 20s. net.
[*In the Press.*

—— POTIPHAR'S WIFE, and other Poems. Crown 8vo. 5s. net.

—— SEAS AND LANDS. With 71 Illustrations. Crown 8vo. 7s. 6d.

—— ADZUMA; OR, THE JAPANESE WIFE. A Play. Cr. 8vo. 6s. 6d. net.

ARNOLD (Dr. T.).—INTRODUCTORY LECTURES ON MODERN HISTORY. 8vo. 7s. 6d.

—— MISCELLANEOUS WORKS. 8vo. 7s. 6d.

ASHLEY (J. W.).—ENGLISH ECONOMIC HISTORY AND THEORY. Part I.—The Middle Ages. Crown 8vo. 5s.

ATELIER (The) du Lys; or, An Art Student in the Reign of Terror. By the Author of 'Mademoiselle Mori'. Crown 8vo. 2s. 6d.

BY THE SAME AUTHOR. Crown 2s. 6d. each.

MADEMOISELLE MORI.
THAT CHILD.
UNDER A CLOUD.
THE FIDDLER OF LUGAU.

A CHILD OF THE REVOLUTION.
HESTER'S VENTURE.
IN THE OLDEN TIME.

—— THE YOUNGER SISTER: a Tale. Crown 8vo. 6s.

BACON.—COMPLETE WORKS. Edited by R. L. ELLIS, J. SPEDDING, and D. D. HEATH. 7 vols. 8vo. £3 13s. 6d.

—— LETTERS AND LIFE, INCLUDING ALL HIS OCCASIONAL WORKS. Edited by J. SPEDDING. 7 vols. 8vo. £4 4s.

—— THE ESSAYS; with Annotations. By Archbishop WHATELY. 8vo. 10s. 6d.

—— THE ESSAYS; with Introduction, Notes, and Index. By E. A. ABBOTT. 2 vols. Fcp. 8vo. 6s. Text and Index only. Fcp. 8vo. 2s. 6d.

BADMINTON LIBRARY (The), edited by the DUKE OF BEAUFORT, assisted by ALFRED E. T. WATSON.

ATHLETICS AND FOOTBALL. By MONTAGUE SHEARMAN. With 41 Illustrations. Crown 8vo. 10s. 6d.

BOATING. By W. B. WOODGATE. With 49 Illustrations. Crown 8vo. 10s. 6d.

COURSING AND FALCONRY. By HARDING COX and the Hon. GERALD LASCELLES. With 76 Illustrations. Crown 8vo. 10s. 6d.

CRICKET. By A. G. STEEL and the Hon. R. H. LYTTELTON. With 63 Illustrations. Crown 8vo. 10s. 6d.

CYCLING. By VISCOUNT BURY (Earl of Albemarle) and G. LACY HILLIER. With 89 Illustrations. Crown 8vo. 10s. 6d.

DRIVING. By the DUKE OF BEAUFORT. With 65 Illustrations. Crown 8vo. 10s. 6d.

FENCING, BOXING, AND WRESTLING. By WALTER H. POLLOCK, F. C. GROVE, C. PREVOST, E. B. MICHELL, and WALTER ARMSTRONG. With 42 Illustrations. Crown 8vo. 10s. 6d.

FISHING. By H. CHOLMONDELEY-PENNELL.
Vol. I. Salmon, Trout, and Grayling. 158 Illustrations. Crown 8vo. 10s. 6d.
Vol. II. Pike and other Coarse Fish. 132 Illustrations. Crown 8vo. 10s. 6d.

[*Continued.*

4   A CATALOGUE OF BOOKS IN GENERAL LITERATURE

BADMINTON LIBRARY (The)—*(continued).*
GOLF. By HORACE HUTCHINSON, the Rt. Hon. A. J. BALFOUR, M.P., ANDREW LANG, Sir W. G. SIMPSON, Bart., &c. With 88 Illustrations. Crown 8vo. 10s. 6d.

HUNTING. By the DUKE OF BEAUFORT, and MOWBRAY MORRIS. With 53 Illustrations. Crown 8vo. 10s. 6d.

MOUNTAINEERING. By C. T. DENT, Sir F. POLLOCK, Bart., W. M. CONWAY, DOUGLAS FRESHFIELD, C. E. MATHEWS, C. PILKINGTON, and other Writers. With Illustrations by H. G. WILLINK.

RACING AND STEEPLECHASING. By the EARL OF SUFFOLK AND BERKSHIRE, W. G. CRAVEN, &c. 56 Illustrations. Crown 8vo. 10s. 6d.

RIDING AND POLO. By Captain ROBERT WEIR, Riding-Master, R.H.G., J. MORAY BROWN, &c. With 59 Illustrations. Crown 8vo. 10s. 6d.

SHOOTING. By LORD WALSINGHAM, and Sir RALPH PAYNE-GALLWEY, Bart.
Vol. I. Field and Covert. With 105 Illustrations. Crown 8vo. 10s. 6d.
Vol. II. Moor and Marsh. With 65 Illustrations. Crown 8vo. 10s. 6d.

SKATING, CURLING, TOBOGGANING, &c. By J. M. HEATHCOTE, C. G. TEBBUTT, T. MAXWELL WITHAM, the Rev. JOHN KERR, ORMOND HAKE, and Colonel BUCK. With 284 Illustrations. Crown 8vo. 10s. 6d.

TENNIS, LAWN TENNIS, RACKETS, AND FIVES. By J. M. and C. G. HEATHCOTE, E. O. PLEYDELL-BOUVERIE, and A. C. AINGER. With 79 Illustrations. Crown 8vo. 10s. 6d.

BAGEHOT (Walter).—BIOGRAPHICAL STUDIES. 8vo. 12s.
——— ECONOMIC STUDIES. 8vo. 10s. 6d.
——— LITERARY STUDIES. 2 vols. 8vo. 28s.
——— THE POSTULATES OF ENGLISH POLITICAL ECONOMY. Crown 8vo. 2s. 6d.

BAGWELL (Richard).—IRELAND UNDER THE TUDORS. (3 vols.) Vols. I. and II. From the first invasion of the Northmen to the year 1578. 8vo. 32s. Vol. III. 1578-1603. 8vo. 18s.

BAIN (Alex.).—MENTAL AND MORAL SCIENCE. Crown 8vo. 10s. 6d.
——— SENSES AND THE INTELLECT. 8vo. 15s.
——— EMOTIONS AND THE WILL. 8vo. 15s.
——— LOGIC, DEDUCTIVE AND INDUCTIVE. Part I., *Deduction*, 4s. Part II., *Induction*, 6s. 6d.
——— PRACTICAL ESSAYS. Crown 8vo. 2s.

BAKER (Sir S. W.).—EIGHT YEARS IN CEYLON. With 6 Illustrations. Crown 8vo. 3s. 6d.
——— THE RIFLE AND THE HOUND IN CEYLON. With 6 Illustrations. Crown 8vo. 3s. 6d.

BALL (The Rt. Hon. T. J.).—THE REFORMED CHURCH OF IRELAND (1537-1889). 8vo. 7s. 6d.
——— HISTORICAL REVIEW OF THE LEGISLATIVE SYSTEMS OPERATIVE IN IRELAND (1172-1800). 8vo. 6s.

BARING-GOULD (Rev. S.).—CURIOUS MYTHS OF THE MIDDLE AGES. Crown 8vo. 3s. 6d.
——— ORIGIN AND DEVELOPMENT OF RELIGIOUS BELIEF. 2 vols. Crown 8vo. 3s. 6d. each.

BEACONSFIELD (The Earl of).—NOVELS AND TALES. The Hughenden Edition. With 2 Portraits and 11 Vignettes. 11 vols. Crown 8vo. 42s.

Endymion.	Venetia.	Alroy, Ixion, &c.
Lothair.	Henrietta Temple.	The Young Duke, &c.
Coningsby.	Contarini Fleming, &c.	Vivian Grey.
Tancred. Sybil.		

NOVELS AND TALES. Cheap Edition. 11 vols. Crown 8vo. 1s. each, boards; 1s. 6d. each, cloth.

BECKER (Professor).—GALLUS; or, Roman Scenes in the Time of Augustus. Illustrated. Post 8vo. 7s. 6d.

—— CHARICLES; or, Illustrations of the Private Life of the Ancient Greeks. Illustrated. Post 8vo. 7s. 6d.

BELL (Mrs. Hugh).—CHAMBER COMEDIES. Crown 8vo. 6s.

—— NURSERY COMEDIES. Fcp. 8vo. 1s. 6d.

BENT (J. Theodore).—THE RUINED CITIES OF MASHONALAND: being a Record of Excavations and Explorations, 1891-2. With numerous Illustrations and Maps. 8vo. 18s.

BRASSEY (Lady).—A VOYAGE IN THE 'SUNBEAM,' OUR HOME ON THE OCEAN FOR ELEVEN MONTHS.
Library Edition. With 8 Maps and Charts, and 118 Illustrations, 8vo. 21s.
Cabinet Edition. With Map and 66 Illustrations, Crown 8vo. 7s. 6d.
'Silver Library' Edition. With 66 Illustrations, Crown 8vo. 3s. 6d.
School Edition. With 37 Illustrations, Fcp. 2s. cloth, or 3s. white parchment.
Popular Edition. With 60 Illustrations, 4to. 6d. sewed, 1s. cloth.

—— SUNSHINE AND STORM IN THE EAST.
Library Edition. With 2 Maps and 114 Illustrations, 8vo. 21s.
Cabinet Edition. With 2 Maps and 114 Illustrations, Crown 8vo. 7s. 6d.
Popular Edition. With 103 Illustrations, 4to. 6d. sewed, 1s. cloth.

—— IN THE TRADES, THE TROPICS, AND THE 'ROARING FORTIES'.
Cabinet Edition. With Map and 220 Illustrations, Crown 8vo. 7s. 6d.
Popular Edition. With 183 Illustrations, 4to. 6d. sewed, 1s. cloth.

—— THE LAST VOYAGE TO INDIA AND AUSTRALIA IN THE 'SUNBEAM'. With Charts and Maps, and 40 Illustrations in Monotone (20 full-page), and nearly 200 Illustrations in the Text. 8vo. 21s.

—— THREE VOYAGES IN THE 'SUNBEAM'. Popular Edition. With 346 Illustrations, 4to. 2s. 6d.

"BRENDA."—WITHOUT A REFERENCE. A Story for Children. Crown 8vo. 3s. 6d.

—— OLD ENGLAND'S STORY. In little Words for little Children. With 29 Illustrations by SIDNEY P. HALL, &c. Imperial 16mo. 3s. 6d.

BRIGHT (Rev. J. Franck).—A HISTORY OF ENGLAND. 4 vols. Cr. 8vo.
Period I.—Mediæval Monarchy: The Departure of the Romans to Richard III. From A.D. 449 to 1485. 4s. 6d.
Period II.—Personal Monarchy: Henry VII. to James II. From 1485 to 1688. 5s.
Period III.—Constitutional Monarchy: William and Mary to William IV. From 1689 to 1837. 7s. 6d.
Period IV.—The Growth of Democracy: Victoria. From 1837 to 1880. 6s.

BUCKLE (Henry Thomas).—HISTORY OF CIVILISATION IN ENGLAND AND FRANCE, SPAIN AND SCOTLAND. 3 vols. Cr. 8vo. 24s.

BULL (Thomas).—HINTS TO MOTHERS ON THE MANAGEMENT OF THEIR HEALTH during the Period of Pregnancy. Fcp. 8vo. 1s. 6d.

————— THE MATERNAL MANAGEMENT OF CHILDREN IN HEALTH AND DISEASE. Fcp. 8vo. 1s. 6d.

BUTLER (Samuel).—EREWHON. Crown 8vo. 5s.

————— THE FAIR HAVEN. A Work in Defence of the Miraculous Element in our Lord's Ministry. Crown 8vo. 7s. 6d.

————— LIFE AND HABIT. An Essay after a Completer View of Evolution. Cr. 8vo. 7s. 6d.

————— EVOLUTION, OLD AND NEW. Crown 8vo. 10s. 6d.

————— UNCONSCIOUS MEMORY. Crown 8vo. 7s. 6d.

————— ALPS AND SANCTUARIES OF PIEDMONT AND THE CANTON TICINO. Illustrated. Pott 4to. 10s. 6d.

————— SELECTIONS FROM WORKS. Crown 8vo. 7s. 6d.

————— LUCK, OR CUNNING, AS THE MAIN MEANS OF ORGANIC MODIFICATION? Crown 8vo. 7s. 6d.

————— EX VOTO. An Account of the Sacro Monte or New Jerusalem at Varallo-Sesia. Crown 8vo. 10s. 6d.

————— HOLBEIN'S 'LA DANSE'. 3s.

CARLYLE (Thomas).—THOMAS CARLYLE: a History of his Life. By J. A. FROUDE. 1795-1835, 2 vols. Cr. 8vo. 7s. 1834-1881, 2 vols. Cr. 8vo. 7s.

LAST WORDS OF THOMAS CARLYLE—Wotton Reinfred—Excursion (Futile enough) to Paris—Letters to Varnhagen von Ense, &c. Cr. 8vo. 6s. 6d. net.

CHETWYND (Sir George).—RACING REMINISCENCES AND EXPERIENCES OF THE TURF. 2 vols. 8vo. 21s.

CHILD (Gilbert W.).—CHURCH AND STATE UNDER THE TUDORS. 8vo. 15s.

CHILTON (E.).—THE HISTORY OF A FAILURE, and other Tales. Fcp. 8vo. 3s. 6d.

CHISHOLM (G. G.).—HANDBOOK OF COMMERCIAL GEOGRAPHY. New Edition. With 29 Maps. 8vo. 10s. net.

CLERKE (Agnes M.).—FAMILIAR STUDIES IN HOMER. Crown 8vo. 7s. 6d.

CLODD (Edward).—THE STORY OF CREATION: a Plain Account of Evolution. With 77 Illustrations. Crown 8vo. 3s. 6d.

CLUTTERBUCK (W. J.).—ABOUT CEYLON AND BORNEO. With 47 Illustrations. Crown 8vo. 10s. 6d.

COLENSO (J. W.).—THE PENTATEUCH AND BOOK OF JOSHUA CRITICALLY EXAMINED. Crown 8vo. 6s.

COMYN (L. N.).—ATHERSTONE PRIORY: a Tale. Crown 8vo. 2s. 6d.

CONINGTON (John).—THE ÆNEID OF VIRGIL. Translated into English Verse. Crown 8vo. 6s.

————— THE POEMS OF VIRGIL. Translated into English Prose. Cr. 8vo. 6s.

COPLESTON (Reginald Stephen, D.D., Bishop of Colombo).—BUDDHISM, PRIMITIVE AND PRESENT, IN MAGADHA AND IN CEYLON. 8vo. 16s.

COX (Rev. Sir G. W.).—A HISTORY OF GREECE, from the Earliest Period to the Death of Alexander the Great. With 11 Maps. Cr. 8vo. 7s. 6d.

CRAKE (Rev. A. D.).—HISTORICAL TALES. Cr. 8vo. 5 vols. 2s. 6d. each.

Edwy the Fair; or, The First Chronicle of Æscendune.
Alfgar the Dane; or, The Second Chronicle of Æscendune.
The Rival Heirs: being the Third and Last Chronicle of Æscendune.
The House of Walderne. A Tale of the Cloister and the Forest in the Days of the Barons' Wars.
Brian Fitz-Count. A Story of Wallingford Castle and Dorchester Abbey.

———— HISTORY OF THE CHURCH UNDER THE ROMAN EMPIRE, A.D. 30-476. Crown 8vo. 7s. 6d.

CREIGHTON (Mandell, D.D.)—HISTORY OF THE PAPACY DURING THE REFORMATION. 8vo. Vols. I. and II., 1378-1464, 32s.; Vols. III. and IV., 1464-1518, 24s.

CROZIER (John Beattie, M.D.).—CIVILISATION AND PROGRESS. Revised and Enlarged, and with New Preface. More fully explaining the nature of the New Organon used in the solution of its problems. 8vo. 14s.

CRUMP (A.).—A SHORT ENQUIRY INTO THE FORMATION OF POLITICAL OPINION, from the Reign of the Great Families to the Advent of Democracy. 8vo. 7s. 6d.

———— AN INVESTIGATION INTO THE CAUSES OF THE GREAT FALL IN PRICES which took place coincidently with the Demonetisation of Silver by Germany. 8vo. 6s.

CURZON (George N., M.P.).—PERSIA AND THE PERSIAN QUESTION. With 9 Maps, 96 Illustrations, Appendices, and an Index. 2 vols. 8vo. 42s.

DANTE.—LA COMMEDIA DI DANTE. A New Text, carefully Revised with the aid of the most recent Editions and Collations. Small 8vo. 6s.

DE LA SAUSSAYE (Prof. Chantepie).—A MANUAL OF THE SCIENCE OF RELIGION. Translated by Mrs. COLYER FERGUSSON (née MAX MÜLLER). Crown 8vo. 12s. 6d.

DEAD SHOT (THE); or, Sportman's Complete Guide. Being a Treatise on the Use of the Gun, with Rudimentary and Finishing Lessons on the Art of Shooting Game of all kinds, also Game Driving, Wild-Fowl and Pigeon Shooting, Dog Breaking, &c. By MARKSMAN. Crown 8vo. 10s. 6d.

DELAND (Margaret, Author of 'John Ward').—THE STORY OF A CHILD. Crown 8vo. 5s.

DE SALIS (Mrs.).—Works by:—

Cakes and Confections à la Mode. Fcp. 8vo. 1s. 6d.
Dressed Game and Poultry à la Mode. Fcp. 8vo. 1s. 6d.
Dressed Vegetables à la Mode. Fcp. 8vo. 1s. 6d.
Drinks à la Mode. Fcp. 8vo. 1s. 6d.
Entrées à la Mode. Fcp. 1s. 8vo. 6d.
Floral Decorations. Fcp. 8vo. 1s. 6d.
Oysters à la Mode. Fcp. 8vo. 1s. 6d.
Puddings and Pastry à la Mode. Fcp. 8vo. 1s. 6d.

Savouries à la Mode. Fcp. 8vo. 1s. 6d.
Soups and Dressed Fish à la Mode. Fcp. 8vo. 1s. 6d.
Sweets and Supper Dishes à la Mode. Fcp. 8vo. 1s. 6d.
Tempting Dishes for Small Incomes. Fcp. 8vo. 1s. 6d.
Wrinkles and Notions for every Household. Crown 8vo. 1s. 6d.
New-Laid Eggs: Hints for Amateur Poultry Rearers. Fcp. 8vo. 1s. 6d.

DE TOCQUEVILLE (Alexis).—DEMOCRACY IN AMERICA Translated by HENRY REEVE, C.B. 2 vols. Crown 8vo. 16s.

DOROTHY WALLIS: an Autobiography. With Preface by WALTER BESANT. Crown 8vo. 6s.

DOUGALL (L.).—BEGGARS ALL; a Novel. Crown 8vo. 3s. 6d.

DOWELL (Stephen).—A HISTORY OF TAXATION AND TAXES IN ENGLAND. 4 vols. 8vo. Vols. I. and II., The History of Taxation, 21s. Vols. III. and IV., The History of Taxes, 21s.

DOYLE (A. Conan).—MICAH CLARKE: a Tale of Monmouth's Rebellion. With Frontispiece and Vignette. Crown 8vo. 3s. 6d.

—— THE CAPTAIN OF THE POLESTAR; and other Tales. Cr. 8vo. 3s. 6d.

EWALD (Heinrich).—THE ANTIQUITIES OF ISRAEL. 8vo. 12s. 6d.

—— THE HISTORY OF ISRAEL. 8vo. Vols. I. and II. 24s. Vols. III. and IV. 21s. Vol. V. 18s. Vol. VI. 16s. Vol. VII. 21s. Vol. VIII. 18s.

FALKENER (Edward).—GAMES, ANCIENT AND ORIENTAL, AND HOW TO PLAY THEM. Being the Games of the Ancient Egyptians, the Hiera Gramme of the Greeks, the Ludus Latrunculorum of the Romans, and the Oriental Games of Chess, Draughts, Backgammon, and Magic Squares. With numerous Photographs, Diagrams, &c. 8vo. 21s.

FARNELL (G. S.).—GREEK LYRIC POETRY. 8vo. 16s.

FARRAR (F. W.).—LANGUAGE AND LANGUAGES. Crown 8vo. 6s.

—— DARKNESS AND DAWN; or, Scenes in the Days of Nero. An Historic Tale. Crown 8vo. 7s. 6d.

FITZPATRICK (W. J.).—SECRET SERVICE UNDER PITT. 8vo. 14s.

FITZWYGRAM (Major-General Sir F.).—HORSES AND STABLES. With 19 pages of Illustrations. 8vo. 5s.

FORD (Horace).—THE THEORY AND PRACTICE OF ARCHERY. New Edition, thoroughly Revised and Re-written by W. BUTT. 8vo. 14s.

FOUARD (Abbé Constant).—THE CHRIST THE SON OF GOD. With Introduction by Cardinal Manning 2 vols. Crown 8vo. 14s.

—— ST. PETER AND THE FIRST YEARS OF CHRISTIANITY. Translated from the Second Edition, with the Author's sanction, by GEORGE F. X. GRIFFITH. With an Introduction by Cardinal GIBBONS. Cr. 8vo. 9s.

FOX (C. J.).—THE EARLY HISTORY OF CHARLES JAMES FOX. By the Right Hon. Sir. G. O. TREVELYAN, Bart.
Library Edition. 8vo. 18s. | Cabinet Edition. Crown 8vo. 6s.

FRANCIS (Francis).—A BOOK ON ANGLING: including full Illustrated Lists of Salmon Flies. Post 8vo. 15s.

FREEMAN (E. A.).—THE HISTORICAL GEOGRAPHY OF EUROPE. With 65 Maps. 2 vols. 8vo. 31s. 6d.

FROUDE (James A.).—THE HISTORY OF ENGLAND, from the Fall of Wolsey to the Defeat of the Spanish Armada. 12 vols. Crown 8vo. £2 2s.

—— THE DIVORCE OF CATHERINE OF ARAGON: The Story as told by the Imperial Ambassadors resident at the Court of Henry VIII. *In Usum Laicorum*. Crown 8vo. 6s.

—— THE ENGLISH IN IRELAND IN THE EIGHTEENTH CENTURY. 3 vols. Crown 8vo. 18s.

—— SHORT STUDIES ON GREAT SUBJECTS.
Cabinet Edition. 4 vols. Cr. 8vo. 24s. | Cheap Edit. 4 vols. Cr. 8vo. 3s. 6d. ea.

—— THE SPANISH STORY OF THE ARMADA, and other Essays, Historical and Descriptive. Crown 8vo. 6s. *[Continued.*

FROUDE (James A.)—(*Continued*).
——— CÆSAR : a Sketch. Crown 8vo. 3s. 6d.
——— OCEANA ; OR, ENGLAND AND HER COLONIES. With 9 Illustrations. Crown 8vo. 2s. boards, 2s. 6d. cloth.
——— THE ENGLISH IN THE WEST INDIES; or, the Bow of Ulysses. With 9 Illustrations. Crown 8vo. 2s. boards, 2s. 6d. cloth.
——— THE TWO CHIEFS OF DUNBOY; an Irish Romance of the Last Century. Crown 8vo. 3s. 6d.
——— THOMAS CARLYLE, a History of his Life. 1795 to 1835. 2 vols. Crown 8vo. 7s. 1834 to 1881. 2 vols. Crown 8vo. 7s.

GALLWEY (Sir Ralph Payne-).—LETTERS TO YOUNG SHOOTERS. First Series. Crown 8vo. 7s. 6d. Second Series. Crown 8vo. 12s. 6d.

GARDINER (Samuel Rawson).—HISTORY OF ENGLAND, 1603-1642. 10 vols. Crown 8vo. price 6s. each.
——— A HISTORY OF THE GREAT CIVIL WAR, 1642-1649. (3 vols.) Vol. I. 1642-1644. With 24 Maps. 8vo. (*out of print*). Vol. II. 1644-1647. With 21 Maps. 8vo. 24s. Vol. III. 1647-1649. With 8 Maps. 28s.
——— THE STUDENT'S HISTORY OF ENGLAND. Vol. I. B.C. 55-A.D. 1509, with 173 Illustrations, Crown 8vo. 4s. Vol. II. 1509-1689, with 96 Illustrations. Crown 8vo. 4s. Vol. III. 1689-1885, with 109 Illustrations. Crown 8vo. 4s. Complete in 1 vol. With 378 Illustrations. Crown 8vo. 12s.
——— A SCHOOL ATLAS OF ENGLISH HISTORY. A Companion Atlas to 'Student's History of England'. 66 Maps and 22 Plans. Fcap. 4to. 5s.

GOETHE.—FAUST. A New Translation chiefly in Blank Verse; with Introduction and Notes. By JAMES ADEY BIRDS. Crown 8vo. 6s.
——— FAUST. The Second Part. A New Translation in Verse. By JAMES ADEY BIRDS. Crown 8vo. 6s.

GREEN (T. H.)—THE WORKS OF THOMAS HILL GREEN. (3 Vols.) Vols. I. and II. 8vo. 16s. each. Vol. III. 8vo. 21s.
——— THE WITNESS OF GOD AND FAITH : Two Lay Sermons. Fcp. 8vo. 2s.

GREVILLE (C. C. F.).—A JOURNAL OF THE REIGNS OF KING GEORGE IV., KING WILLIAM IV., AND QUEEN VICTORIA. Edited by H. REEVE. 8 vols. Crown 8vo. 6s. each.

GWILT (Joseph).—AN ENCYCLOPÆDIA OF ARCHITECTURE. With more than 1700 Engravings on Wood. 8vo. 52s. 6d.

HAGGARD (H. Rider).—SHE. With 32 Illustrations. Crown 8vo. 3s. 6d.
——— ALLAN QUATERMAIN. With 31 Illustrations. Crown 8vo. 3s. 6d.
——— MAIWA'S REVENGE. Crown 8vo. 1s. boards, 1s. 6d. cloth.
——— COLONEL QUARITCH, V.C. Crown 8vo. 3s. 6d.
——— CLEOPATRA : With 29 Illustrations. Crown 8vo. 3s. 6d.
——— BEATRICE. Crown 8vo. 3s. 6d.
——— ERIC BRIGHTEYES. With 51 Illustrations. Crown 8vo. 6s.
——— NADA THE LILY. With 23 Illustrations by C. H. M. KERR. Cr. 8vo. 6s.

HAGGARD (H. Rider) and LANG (Andrew).—THE WORLD'S DESIRE. Crown 8vo. 6s.

HALLIWELL-PHILLIPPS (J. O.)—A CALENDAR OF THE HALLIWELL-PHILLIPPS COLLECTION OF SHAKESPEAREAN RARITIES. Second Edition. Enlarged by Ernest E. Baker. 8vo. 10s. 6d.

———— OUTLINES OF THE LIFE OF SHAKESPEARE. With numerous Illustrations and Facsimiles. 2 vols. Royal 8vo. 21s.

HARRISON (Jane E.).—MYTHS OF THE ODYSSEY IN ART AND LITERATURE. Illustrated with Outline Drawings. 8vo. 18s.

HARRISON (Mary).—COOKERY FOR BUSY LIVES AND SMALL INCOMES. Fcp. 8vo. 1s.

HARTE (Bret).—IN THE CARQUINEZ WOODS. Fcp. 8vo. 1s. bds., 1s. 6d. cloth.

———— BY SHORE AND SEDGE. 16mo. 1s.

———— ON THE FRONTIER. 16mo. 1s.

₊ Complete in one Volume. Crown 8vo. 3s. 6d.

HARTWIG (Dr.).—THE SEA AND ITS LIVING WONDERS. With 12 Plates and 303 Woodcuts. 8vo. 7s. net.

THE TROPICAL WORLD. With 8 Plates and 172 Woodcuts. 8vo. 7s. net.

THE POLAR WORLD. With 3 Maps, 8 Plates and 85 Woodcuts. 8vo. 7s. net.

THE SUBTERRANEAN WORLD. With 3 Maps and 80 Woodcuts. 8vo. 7s. net.

THE AERIAL WORLD. With Map, 8 Plates and 60 Woodcuts. 8vo. 7s. net.

HAVELOCK.—MEMOIRS OF SIR HENRY HAVELOCK, K.C.B. By JOHN CLARK MARSHMAN. Crown 8vo. 3s. 6d.

HEARN (W. Edward).—THE GOVERNMENT OF ENGLAND: its Structure and its Development. 8vo. 16s.

———— THE ARYAN HOUSEHOLD: its Structure and ts Development. An Introduction to Comparative Jurisprudence. 8vo. 16s.

HISTORIC TOWNS. Edited by E. A. FREEMAN and Rev. WILLIAM HUNT. With Maps and Plans. Crown 8vo 3s. 6d. each.

Bristol. By Rev. W. Hunt.
Carlisle. By Dr. Mandell Creighton.
Cinque Ports. By Montagu Burrows.
Colchester. By Rev. E. L. Cutts.
Exeter. By E. A. Freeman.
London. By Rev. W. J. Loftie.

Oxford. By Rev. C. W. Boase.
Winchester. By Rev. G. W. Kitchin.
New York. By Theodore Roosevelt.
Boston (U.S.). By Henry Cabot Lodge.
York. By Rev. James Raine.

HODGSON (Shadworth H.).—TIME AND SPACE: a Metaphysical Essay. 8vo. 16s.

———— THE THEORY OF PRACTICE: an Ethical Enquiry. 2 vols. 8vo. 24s.

———— THE PHILOSOPHY OF REFLECTION. 2 vols. 8vo. 21s.

———— OUTCAST ESSAYS AND VERSE TRANSLATIONS. Crown 8vo. 8s. 6d.

HOOPER (George).—ABRAHAM FABERT: Governor of Sedan, Marshall of France. His Life and Times, 1599-1662. With a Portrait. 8vo. 10s. 6d.

HOWITT (William).—VISITS TO REMARKABLE PLACES. 80 Illustrations. Crown 8vo. 3s. 6d.

HULLAH (John).—COURSE OF LECTURES ON THE HISTORY OF MODERN MUSIC. 8vo. 8s. 6d.

——— COURSE OF LECTURES ON THE TRANSITION PERIOD OF MUSICAL HISTORY. 8vo. 10s. 6d.

HUME.—THE PHILOSOPHICAL WORKS OF DAVID HUME. Edited by T. H. GREEN and T. H. GROSE. 4 vols. 8vo. 56s.

HUTH (Alfred H.).—THE MARRIAGE OF NEAR KIN, considered with respect to the Law of Nations, the Result of Experience, and the Teachings of Biology. Royal 8vo. 21s.

INGELOW (Jean).—POETICAL WORKS. 2 vols. Fcp. 8vo. 12s.

——— LYRICAL AND OTHER POEMS. Selected from the Writings of JEAN INGELOW. Fcp. 8vo. 2s. 6d. cloth plain, 3s. cloth gilt.

——— VERY YOUNG and QUITE ANOTHER STORY: Two Stories. Crown 8vo. 6s.

INGRAM (T. Dunbar).—ENGLAND AND ROME: a History of the Relations between the Papacy and the English State and Church from the Norman Conquest to the Revolution of 1688. 8vo. 14s.

JAMESON (Mrs.).—SACRED AND LEGENDARY ART. With 19 Etchings and 187 Woodcuts. 2 vols. 8vo. 20s. net.

——— LEGENDS OF THE MADONNA, the Virgin Mary as represented in Sacred and Legendary Art. With 27 Etchings and 165 Woodcuts. 8vo. 10s. net.

——— LEGENDS OF THE MONASTIC ORDERS. With 11 Etchings and 88 Woodcuts. 8vo. 10s. net.

——— HISTORY OF OUR LORD. His Types and Precursors. Completed by LADY EASTLAKE. With 31 Etchings and 281 Woodcuts. 2 vols. 8vo. 20s. net.

JEFFERIES (Richard).—FIELD AND HEDGEROW. Last Essays. Crown 8vo. 3s. 6d.

——— THE STORY OF MY HEART: My Autobiography. Crown 8vo. 3s. 6d.

——— RED DEER. With 17 Illustrations by J. CHARLTON and H. TUNALY. Crown 8vo. 3s. 6d.

——— THE TOILERS OF THE FIELD. With autotype reproduction of bust of Richard Jefferies in Salisbury Cathedral. Crown 8vo. 6s.

JENNINGS (Rev. A. C.).—ECCLESIA ANGLICANA. A History of the Church of Christ in England. Crown 8vo. 7s. 6d.

JEWSBURY.—A SELECTION FROM THE LETTERS OF GERALDINE JEWSBURY TO JANE WELSH CARLYLE. Edited by Mrs. ALEXANDER IRELAND, and Prefaced by a Monograph on Miss Jewsbury by the Editor. 8vo. 6s.

JOHNSON (J. & J. H.).—THE PATENTEE'S MANUAL; a Treatise on the Law and Practice of Letters Patent. 8vo. 10s. 6d.

JORDAN (William Leighton).—THE STANDARD OF VALUE. 8vo. 6s.

JUSTINIAN.—THE INSTITUTES OF JUSTINIAN; Latin Text, with English Introduction, &c. By THOMAS C. SANDARS. 8vo. 18s.

KANT (Immanuel).—CRITIQUE OF PRACTICAL REASON, AND OTHER WORKS ON THE THEORY OF ETHICS. 8vo. 12s. 6d.

——— INTRODUCTION TO LOGIC. Translated by T. K. Abbott. Notes by S. T. Coleridge. 8vo. 6s.

KEITH DERAMORE. A Novel. By the Author of 'Miss Molly'. Cr. 8vo. 6s.

KILLICK (Rev. A. H.).—HANDBOOK TO MILL'S SYSTEM OF LOGIC. Crown 8vo. 3s. 6d.

KNIGHT (E. F.).—THE CRUISE OF THE 'ALERTE'; the Narrative of a Search for Treasure on the Desert Island of Trinidad. With 2 Maps and 23 Illustrations. Crown 8vo. 3s. 6d.

——— WHERE THREE EMPIRES MEET. A Narrative of Recent Travel in Kashmir, Western Tibet, Gilgit, and the adjacent countries. 8vo.

LADD (George T.).—ELEMENTS OF PHYSIOLOGICAL PSYCHOLOGY. 8vo. 21s.

——— OUTLINES OF PHYSIOLOGICAL PSYCHOLOGY. A Text-Book of Mental Science for Academies and Colleges. 8vo. 12s.

LANG (Andrew).—CUSTOM AND MYTH: Studies of Early Usage and Belief. With 15 Illustrations. Crown 8vo. 7s. 6d.

——— HOMER AND THE EPIC. Crown 8vo. 9s. net.

——— BOOKS AND BOOKMEN. With 2 Coloured Plates and 17 Illustrations. Fcp. 8vo. 2s. 6d. net.

——— LETTERS TO DEAD AUTHORS. Fcp. 8vo. 2s. 6d. net.
——— OLD FRIENDS. Fcp. 8vo. 2s. 6d. net.
——— LETTERS ON LITERATURE. Fcp. 8vo. 2s. 6d. net.
——— GRASS OF PARNASSUS. Fcp. 8vo. 2s. 6d. net.
——— BALLADS OF BOOKS. Edited by ANDREW LANG. Fcp. 8vo. 6s.
——— THE BLUE FAIRY BOOK. Edited by ANDREW LANG. With 8 Plates and 130 Illustrations in the Text. Crown 8vo. 6s.
——— THE RED FAIRY BOOK. Edited by ANDREW LANG. With 4 Plates and 96 Illustrations in the Text. Crown 8vo. 6s.
——— THE BLUE POETRY BOOK. With 12 Plates and 88 Illustrations in the Text. Crown 8vo. 6s.
——— THE BLUE POETRY BOOK. School Edition, without Illustrations. Fcp. 8vo. 2s. 6d.
——— THE GREEN FAIRY BOOK. Edited by ANDREW LANG. With 13 Plates and 88 Illustrations in the Text by H. J. Ford. Crown 8vo. 6s.
——— ANGLING SKETCHES. With Illustrations by W. G. BURN-MURDOCH. Crown 8vo. 7s. 6d.

LAVISSE (Ernest).—GENERAL VIEW OF THE POLITICAL HISTORY OF EUROPE. Crown 8vo. 5s.

LECKY (W. E. H.).—HISTORY OF ENGLAND IN THE EIGHTEENTH CENTURY. Library Edition, 8vo. Vols. I. and II. 1700-1760. 36s. Vols. III. and IV. 1760-1784. 36s. Vols. V. and VI. 1784-1793. 36s. Vols. VII. and VIII. 1793-1800. 36s. Cabinet Edition, 12 vols. Crown 8vo. 6s. each.

——— THE HISTORY OF EUROPEAN MORALS FROM AUGUSTUS TO CHARLEMAGNE. 2 vols. Crown 8vo. 16s.

——— HISTORY OF THE RISE AND INFLUENCE OF THE SPIRIT OF RATIONALISM IN EUROPE. 2 vols. Crown 8vo. 16s.

——— POEMS. Fcap. 8vo. 5s.

LEES (J. A.) and CLUTTERBUCK (W. J.).—B.C. 1887, A RAMBLE IN BRITISH COLUMBIA. With Map and 75 Illusts. Cr. 8vo. 3s. 6d.

LEWES (George Henry).—THE HISTORY OF PHILOSOPHY, from Thales to Comte. 2 vo s. 8vo. 32s.

LEYTON (Frank).—THE SHADOWS OF THE LAKE, and other Poems. Crown 8vo. 7s. 6d. Cheap Edition. Crown 8vo. 3s. 6d.

LLOYD (F. J.).—THE SCIENCE OF AGRICULTURE. 8vo. 12s.

LONGMAN (Frederick W.).—CHESS OPENINGS. Fcp. 8vo. 2s. 6d.

———— FREDERICK THE GREAT AND THE SEVEN YEARS' WAR. Fcp. 8vo. 2s. 6d.

LONGMORE (Sir T.).—RICHARD WISEMAN, Surgeon and Sergeant-Surgeon to Charles II. A Biographical Study. With Portrait. 8vo. 10s. 6d.

LOUDON (J. C.).—ENCYCLOPÆDIA OF GARDENING. With 1000 Woodcuts. 8vo. 21s.

———— ENCYCLOPÆDIA OF AGRICULTURE; the Laying-out, Improvement, and Management of Landed Property. With 1100 Woodcuts. 8vo. 21s.

———— ENCYCLOPÆDIA OF PLANTS; the Specific Character, &c., of all Plants found in Great Britain. With 12,000 Woodcuts. 8vo. 42s.

LUBBOCK (Sir J.).—THE ORIGIN OF CIVILISATION and the Primitive Condition of Man. With 5 Plates and 20 Illustrations in the Text. 8vo. 18s.

LYALL (Edna).—THE AUTOBIOGRAPHY OF A SLANDER. Fcp. 8vo. 1s. sewed.

Presentation Edition, with 20 Illustrations by L. SPEED. Crown 8vo. 5s.

LYDEKKER (R., B.A.).—PHASES OF ANIMAL LIFE, PAST AND PRESENT. With 82 Illustrations. Crown 8vo. 6s.

LYDE (Lionel W.).—AN INTRODUCTION TO ANCIENT HISTORY. With 3 Coloured Maps. Crown 8vo. 3s.

LYONS (Rev. Daniel).—CHRISTIANITY AND INFALLIBILITY— Both or Neither. Crown 8vo. 5s.

LYTTON (Earl of).—MARAH.—By OWEN MEREDITH (the late Earl of Lytton). Fcp. 8vo. 6s. 6d.

———— KING POPPY; a Fantasia. Crown 8vo. 10s. 6d.

MACAULAY (Lord).—COMPLETE WORKS OF LORD MACAULAY. Library Edition, 8 vols. 8vo. £5 5s. | Cabinet Edition, 16 vols. post 8vo. £4 16s.

———— HISTORY OF ENGLAND FROM THE ACCESSION OF JAMES THE SECOND.

Popular Edition, 2 vols. Crown 8vo. 5s. | People's Edition, 4 vols. Crown 8vo. 16s.
Student's Edition, 2 vols. Crown 8vo. | Cabinet Edition, 8 vols. Post 8vo. 48s.
12s. | Library Edition, 5 vols. 8vo. £4.

———— CRITICAL AND HISTORICAL ESSAYS, WITH LAYS OF ANCIENT ROME, in 1 volume.

Popular Edition, Crown 8vo. 2s. 6d. | 'Silver Library' Edition. With Portrait and Illustrations to the 'Lays'.
Authorised Edition, Crown 8vo. 2s. 6d., or 3s. 6d. gilt edges. | Crown 8vo. 3s. 6d.

———— CRITICAL AND HISTORICAL ESSAYS.

Student's Edition. Crown 8vo. 6s. | Trevelyan Edition, 2 vols. Crown 8vo. 9s.
People's Edition, 2 vols. Crown 8vo. 8s. | Cabinet Edition, 4 vols. Post 8vo. 24s.
| Library Edition, 3 vols. 8vo. 36s.

[Continued.

MACAULAY (Lord)—(*Continued*).

—— ESSAYS which may be had separately, price 6d. each sewed. 1s. each cloth.

Addison and Walpole.
Frederic the Great.
Croker's Boswell's Johnson.
Hallam's Constitutional History.
Warren Hastings (3d. sewed, 6d. cloth).
The Earl of Chatham (Two Essays).

Ranke and Gladstone.
Milton and Machiavelli.
Lord Bacon.
Lord Clive.
Lord Byron, and the Comic Dramatists of the Restoration.

The Essay on Warren Hastings, annotated by S. Hales. Fcp. 8vo. 1s. 6d.

The Essay on Lord Clive, annotated by H. Courthope Bowen. Fcp. 8vo. 2s. 6d.

—— SPEECHES. People's Edition, Crown 8vo. 3s. 6d.

—— LAYS OF ANCIENT ROME, &c. Illustrated by G. Scharf. Library Edition. Fcp. 4to. 10s. 6d.

Bijou Edition, 18mo. 2s. 6d. gilt top.

Popular Edition, Fcp. 4to. 6d sewed, 1s. cloth.

—— Illustrated by J. R. Weguelin. Crown 8vo. 3s. 6d. gilt edges.

Cabinet Edition, Post 8vo. 3s. 6d.

Annotated Edition, Fcp. 8vo. 1s. sewed, 1s. 6d. cloth.

—— MISCELLANEOUS WRITINGS.

People's Edition. Crown 8vo. 4s. 6d. | Library Edition, 2 vols. 8vo. 21s.

—— MISCELLANEOUS WRITINGS AND SPEECHES.

Popular Edition. Crown 8vo. 2s. 6d. | Cabinet Edition, Post 8vo. 24s.
Student's Edition. Crown 8vo. 6s.

—— SELECTIONS FROM THE WRITINGS OF LORD MACAULAY. Edited, with Notes, by the Right Hon. Sir G. O. TREVELYAN. Crown 8vo. 6s.

—— THE LIFE AND LETTERS OF LORD MACAULAY. By the Right Hon. Sir G. O. TREVELYAN.

Popular Edition. Crown. 8vo. 2s. 6d. | Cabinet Edition, 2 vols. Post 8vo. 12s.
Student's Edition. Crown 8vo. 6s. | Library Edition, 2 vols. 8vo. 36s.

MACDONALD (George).—UNSPOKEN SERMONS. Three Series. Crown 8vo. 3s. 6d. each.

—— THE MIRACLES OF OUR LORD. Crown 8vo. 3s. 6d.

—— A BOOK OF STRIFE, IN THE FORM OF THE DIARY OF AN OLD SOUL: Poems. 12mo. 6s.

MACFARREN (Sir G. A.).—LECTURES ON HARMONY. 8vo. 12s.

MACKAIL (J. W.).—SELECT EPIGRAMS FROM THE GREEK ANTHOLOGY. With a Revised Text, Introduction, Translation, &c. 8vo. 16s.

MACLEOD (Henry D.).—THE ELEMENTS OF BANKING. Crown 8vo. 3s. 6d.

—— THE THEORY AND PRACTICE OF BANKING. Vol. I. 8vo. 12s., Vol. II. 14s.

—— THE THEORY OF CREDIT. 8vo. Vol. I. [*New Edition in the Press*]; Vol. II. Part I. 4s. 6d.; Vol. II. Part II. 10s. 6d.

MANNERING (G. E.).—WITH AXE AND ROPE IN THE NEW ZEALAND ALPS. Illustrated. 8vo. 12s. 6d.

MANUALS OF CATHOLIC PHILOSOPHY (*Stonyhurst Series*).

Logic. By Richard F. Clarke. Crown 8vo. 5*s.*
First Principles of Knowledge. By John Rickaby. Crown 8vo. 5*s.*
Moral Philosophy (Ethics and Natural Law). By Joseph Rickaby. Crown 8vo. 5*s.*
General Metaphysics. By John Rickaby. Crown 8vo. 5*s.*
Psychology. By Michael Maher. Crown 8vo. 6*s.* 6*d.*
Natural Theology. By Bernard Boedder. Crown 8vo. 6*s.* 6*d.*
A Manual of Political Economy. By C. S. Devas. 6*s.* 6*d.*

MARBOT (Baron de).—THE MEMOIRS OF. Translated from the French. Crown 8vo. 7*s.* 6*d.*

MARTINEAU (James).—HOURS OF THOUGHT ON SACRED THINGS. Two Volumes of Sermons. 2 vols. Crown 8vo. 7*s.* 6*d.* each.

——— ENDEAVOURS AFTER THE CHRISTIAN LIFE. Discourses. Crown 8vo. 7*s.* 6*d.*

——— HOME PRAYERS. Crown 8vo. 3*s.* 6*d.*

——— THE SEAT OF AUTHORITY IN RELIGION. 8vo. 14*s.*

——— ESSAYS, REVIEWS, AND ADDRESSES. 4 vols. Crown 8vo. 7*s.* 6*d.* each.

I. Personal: Political.
II. Ecclesiastical: Historical.
III. Theological: Philosophical.
IV. Academical: Religious.

MATTHEWS (Brander).—A FAMILY TREE, and other Stories. Crown 8vo. 6*s.*

——— PEN AND INK—Selected Papers. Crown 8vo. 5*s.*

——— WITH MY FRIENDS: Tales told in Partnership. Crown 8vo. 6*s.*

MAUNDER'S TREASURIES. Fcp. 8vo. 6*s.* each volume

Biographical Treasury.
Treasury of Natural History. With 900 Woodcuts.
Treasury of Geography. With 7 Maps and 16 Plates.
Scientific and Literary Treasury.
Historical Treasury.
Treasury of Knowledge.
The Treasury of Bible Knowledge. By the Rev. J. AYRE. With 5 Maps, 15 Plates, and 300 Woodcuts. Fcp. 8vo. 6*s.*
The Treasury of Botany. Edited by J. LINDLEY and T. MOORE. With 274 Woodcuts and 20 Steel Plates. 2 vols.

MAX MÜLLER (F.).—SELECTED ESSAYS ON LANGUAGE, MYTHOLOGY, AND RELIGION. 2 vols. Crown 8vo. 16*s.*

——— THREE LECTURES ON THE SCIENCE OF LANGUAGE. Cr. 8vo. 3*s.*

——— THE SCIENCE OF LANGUAGE, founded on Lectures delivered at the Royal Institution in 1861 and 1863. 2 vols. Crown 8vo. 21*s.*

——— HIBBERT LECTURES ON THE ORIGIN AND GROWTH OF RELIGION, as illustrated by the Religions of India. Crown 8vo. 7*s.* 6*d.*

——— INTRODUCTION TO THE SCIENCE OF RELIGION; Four Lectures delivered at the Royal Institution. Crown 8vo. 7*s.* 6*d.*

——— NATURAL RELIGION. The Gifford Lectures, delivered before the University of Glasgow in 1888. Crown 8vo. 10*s.* 6*d.*

——— PHYSICAL RELIGION. The Gifford Lectures, delivered before the University of Glasgow in 1890. Crown 8vo. 10*s.* 6*d.*

——— ANTHROPOLOGICAL RELIGION: The Gifford Lectures delivered before the University of Glasgow in 1891. Crown 8vo. 10*s.* 6*d.*

[*Continued.*

MAX MÜLLER (F.)—*(Continued).*

——— THEOSOPHY OR PSYCHOLOGICAL RELIGION: the Gifford Lectures delivered before the University of Glasgow in 1892. Crown 8vo.

——— THE SCIENCE OF THOUGHT. 8vo. 21s.

——— THREE INTRODUCTORY LECTURES ON THE SCIENCE OF THOUGHT. 8vo. 2s. 6d.

——— BIOGRAPHIES OF WORDS, AND THE HOME OF THE ARYAS. Crown 8vo. 7s. 6d.

——— INDIA, WHAT CAN IT TEACH US? Crown 8vo. 3s. 6d.

——— A SANSKRIT GRAMMAR FOR BEGINNERS. New and Abridged Edition. By A. A. MacDonell. Crown 8vo. 6s.

MAY (Sir Thomas Erskine).—THE CONSTITUTIONAL HISTORY OF ENGLAND since the Accession of George III. 3 vols. Crown 8vo. 18s.

MEADE (L. T.).—DADDY'S BOY. With Illustrations. Crown 8vo. 3s. 6d.

——— DEB AND THE DUCHESS. Illust. by M. E. Edwards. Cr. 8vo. 3s. 6d.

——— THE BERESFORD PRIZE. Illustrated by M. E. Edwards. Cr. 8vo. 5s.

MEATH (The Earl of).—SOCIAL ARROWS: Reprinted Articles on various Social Subjects. Crown 8vo. 5s.

——— PROSPERITY OR PAUPERISM? Physical, Industrial, and Technical Training. Edited by the Earl of Meath. 8vo. 5s.

MELVILLE (G. J. Whyte).—Novels by. Crown 8vo. 1s. each, boards; 1s. 6d. each, cloth.

The Gladiators.	The Queen's Maries.	Digby Grand.
The Interpreter.	Holmby House.	General Bounce.
Good for Nothing.	Kate Coventry.	

MENDELSSOHN.—THE LETTERS OF FELIX MENDELSSOHN. Translated by Lady Wallace. 2 vols. Crown 8vo. 10s.

MERIVALE (Rev. Chas.).—HISTORY OF THE ROMANS UNDER THE EMPIRE. Cabinet Edition, 8 vols. Crown 8vo. 48s. Popular Edition, 8 vols. Crown 8vo. 3s. 6d. each.

——— THE FALL OF THE ROMAN REPUBLIC: a Short History of the Last Century of the Commonwealth. 12mo. 7s. 6d.

——— GENERAL HISTORY OF ROME FROM B.C. 753 TO A.D. 476. Cr. 8vo. 7s. 6d.

——— THE ROMAN TRIUMVIRATES. With Maps. Fcp. 8vo. 2s. 6d.

MILL (James).—ANALYSIS OF THE PHENOMENA OF THE HUMAN MIND. 2 vols. 8vo. 28s.

MILL (John Stuart).—PRINCIPLES OF POLITICAL ECONOMY.
Library Edition, 2 vols. 8vo. 30s. | People's Edition, 1 vol. Crown 8vo. 3s. 6d.

——— A SYSTEM OF LOGIC. Crown 8vo. 3s. 6d.

——— ON LIBERTY. Crown 8vo. 1s. 4d.

——— ON REPRESENTATIVE GOVERNMENT. Crown 8vo. 2s.

——— UTILITARIANISM. 8vo. 5s.

——— EXAMINATION OF SIR WILLIAM HAMILTON'S PHILOSOPHY. 8vo. 16s.

——— NATURE, THE UTILITY OF RELIGION AND THEISM. Three Essays, 8vo. 5s.

MILNER (George).—COUNTRY PLEASURES; the Chronicle of a Year chiefly in a Garden. Crown 8vo. 3s. 6d.

MOLESWORTH (Mrs.).—SILVERTHORNS. With Illustrations by F. Noel Paton. Cr. 8vo. 5s.
—— THE PALACE IN THE GARDEN. With Illustrations. Cr. 8vo. 5s.
—— THE THIRD MISS ST. QUENTIN. Crown 8vo. 6s.
—— NEIGHBOURS. With Illustrations by M. Ellen Edwards. Cr. 8vo. 6s.
—— THE STORY OF A SPRING MORNING. With Illustrations. Cr.8vo. 5s.
—— STORIES OF THE SAINTS FOR CHILDREN: the Black Letter Saints. With Illustrations. Royal 16mo. 5s.

MOORE (Edward).—DANTE AND HIS EARLY BIOGRAPHERS. Crown 8vo. 4s. 6d.

MULHALL (Michael G.).—HISTORY OF PRICES SINCE THE YEAR 1850. Crown 8vo. 6s.

NANSEN (Dr. Fridtjof).—THE FIRST CROSSING OF GREENLAND. With numerous Illustrations and a Map. Crown 8vo. 7s. 6d.

NAPIER.—THE LIFE OF SIR JOSEPH NAPIER, BART., EX-LORD CHANCELLOR OF IRELAND. By ALEX. CHARLES EWALD. 8vo. 15s.
—— THE LECTURES, ESSAYS, AND LETTERS OF THE RIGHT HON. SIR JOSEPH NAPIER, BART. 8vo. 12s. 6d.

NESBIT (E.).—LEAVES OF LIFE: Verses. Crown 8vo. 5s.
—— LAYS AND LEGENDS. FIRST Series. Crown 8vo. 3s. 6d. SECOND Series. With Portrait. Crown 8vo. 5s.

NEWMAN (Cardinal).—Works by:—

Discourses to Mixed Congregations. Cabinet Edition, Crown 8vo. 6s. Cheap Edition, 3s. 6d.

Sermons on Various Occasions. Cabinet Edition, Cr. 8vo. 6s. Cheap Edition, 3s. 6d.

The Idea of a University defined and illustrated. Cabinet Edition, Cr. 8vo. 7s. Cheap Edition, Cr. 8vo. 3s. 6d.

Historical Sketches. Cabinet Edition, 3 vols. Crown 8vo. 6s. each. Cheap Edition, 3 vols. Cr. 8vo. 3s. 6d. each.

The Arians of the Fourth Century. Cabinet Edition, Crown 8vo. 6s. Cheap Edition, Crown 8vo. 3s. 6d.

Select Treatises of St. Athanasius in Controversy with the Arians. Freely Translated. 2 vols. Crown 8vo. 15s.

Discussions and Arguments on Various Subjects. Cabinet Edition, Crown 8vo. 6s. Cheap Edition, Crown 8vo. 3s. 6d.

Apologia Pro Vita Sua. Cabinet Ed., Crown 8vo. 6s. Cheap Ed. 3s. 6d.

Development of Christian Doctrine. Cabinet Edition, Crown 8vo. 6s. Cheap Edition, Cr. 8vo. 3s. 6d.

Certain Difficulties felt by Anglicans in Catholic Teaching Considered. Cabinet Edition. Vol. I. Crown 8vo. 7s. 6d.; Vol. II. Crown 8vo. 5s. 6d. Cheap Edition, 2 vols. Crown 8vo. 3s. 6d. each.

The Via Media of the Anglican Church, Illustrated in Lectures, &c. Cabinet Edition, 2 vols. Cr. 8vo. 6s. each. Cheap Edition, 2 vols. Crown 8vo. 3s. 6d. each.

Essays, Critical and Historical. Cabinet Edition, 2 vols. Crown 8vo. 12s. Cheap Edition, 2 vols. Cr. 8vo. 7s.

Biblical and Ecclesiastical Miracles. Cabinet Edition, Crown 8vo. 6s. Cheap Edition, Crown 8vo. 3s. 6d.

Present Position of Catholics in England. Cabinet Edition, Crown 8vo. 7s. 6d. Cheap Edition, Crown 8vo. 3s. 6d.

Tracts. 1. Dissertatiunculæ. 2. On the Text of the Seven Epistles of St. Ignatius. 3. Doctrinal Causes of Arianism. 4. Apollinarianism. 5. St. Cyril's Formula. 6. Ordo de Tempore. 7. Douay Version of Scripture. Crown 8vo. 8s.

[*Continued.*

NEWMAN (Cardinal).—Works by :—(*continued*).

An Essay in Aid of a Grammar of Assent. Cabinet Edition, Crown 8vo. 7s. 6d. Cheap Edition, Crown 8vo. 3s. 6d.

Callista: a Tale of the Third Century. Cabinet Edition, Crown 8vo. 6s. Cheap Edition, Crown 8vo. 3s. 6d.

Loss and Gain: a Tale. Cabinet Edition, Crown 8vo. 6s. Cheap Edition, Crown 8vo. 3s. 6d.

The Dream of Gerontius. 16mo. 6d. sewed, 1s. cloth.

Verses on Various Occasions. Cabinet Edition, Crown 8vo. 6s. Cheap Edition, Crown 8vo. 3s. 6d.

*** *For Cardinal Newman's other Works see Messrs. Longmans & Co.'s Catalogue of Theological Works.*

NORTON (Charles L.).—A HANDBOOK OF FLORIDA. 49 Maps and Plans. Fcp. 8vo. 5s.

O'BRIEN (William).—WHEN WE WERE BOYS: A Novel. Cr. 8vo. 2s. 6d.

OLIPHANT (Mrs.).—MADAM. Crown 8vo. 1s. boards ; 1s. 6d. cloth.

——— IN TRUST. Crown 8vo. 1s. boards; 1s. 6d. cloth.

OMAN (C. W. C.).—A HISTORY OF GREECE FROM THE EARLIEST TIMES TO THE MACEDONIAN CONQUEST. With Maps. Crown 8vo. 4s. 6d.

PARKES (Sir Henry).—FIFTY YEARS IN THE MAKING OF AUSTRALIAN HISTORY. With Portraits. 2 vols. 8vo. 32s.

PAUL (Hermann).—PRINCIPLES OF THE HISTORY OF LANGUAGE. Translated by H. A. Strong. 8vo. 10s. 6d.

PAYN (James).—THE LUCK OF THE DARRELLS. Cr. 8vo. 1s. bds. ; 1s. 6d. cl.

——— THICKER THAN WATER. Crown 8vo. 1s. boards ; 1s. 6d. cloth.

PERRING (Sir Philip).—HARD KNOTS IN SHAKESPEARE. 8vo. 7s. 6d.

——— THE 'WORKS AND DAYS' OF MOSES. Crown 8vo. 3s. 6d.

PHILLIPPS-WOLLEY (C.).—SNAP: a Legend of the Lone Mountain. With 13 Illustrations by H. G. Willink. Crown 8vo. 3s. 6d.

POLE (W.).—THE THEORY OF THE MODERN SCIENTIFIC GAME OF WHIST. Fcp. 8vo. 2s. 6d.

POOLE (W. H. and Mrs.).—COOKERY FOR THE DIABETIC. Fcp. 8vo. 2s. 6d.

PRAEGER (F.).—WAGNER AS I KNEW HIM. Crown 8vo. 7s. 6d.

PRATT (A. E., F.R.G.S.).—TO THE SNOWS OF TIBET THROUGH CHINA. With 33 Illustrations and a Map. 8vo. 18s.

PRENDERGAST (John P.).—IRELAND, FROM THE RESTORATION TO THE REVOLUTION, 1660-1690. 8vo. 5s.

PROCTOR (R.A.).—Works by :—
Old and New Astronomy. 4to. 36s.
The Orbs Around Us. Crown 8vo. 5s.
Other Worlds than Ours. With 14 Illustrations. Crown 8vo. 5s. Cheap Edition, 3s. 6d.
The Moon. Crown 8vo. 5s.
Universe of Stars. 8vo. 10s. 6d.
Larger Star Atlas for the Library, in 12 Circular Maps, with Introduction and 2 Index Pages. Folio, 15s. or Maps only, 12s. 6d.
The Student's Atlas. In 12 Circular Maps. 8vo. 5s.
New Star Atlas. In 12 Circular Maps. Crown 8vo. 5s.
Light Science for Leisure Hours. 3 vols. Crown 8vo. 5s. each.
Chance and Luck. Crown 8vo. 2s. boards ; 2s. 6d. cloth.
Pleasant Ways in Science. Cr. 8vo. 5s. Cheap Edition, 3s. 6d.
How to Play Whist : with the Laws and Etiquette of Whist. Crown 8vo. 3s.6d.
Home Whist : an Easy Guide to Correct Play. 16mo. 1s.

The Stars in their Season. 12 Maps. Royal 8vo. 5s.
Star Primer. Showing the Starry Sky Week by Week, in 24 Hourly Maps. Crown 4to. 2s. 6d.
The Seasons Pictured in 48 Sun-Views of the Earth, and 24 Zodiacal Maps, &c. Demy 4to. 5s.
Strength and Happiness. With 9 Illustrations. Crown 8vo. 5s.
Strength : How to get Strong and keep Strong. Crown 8vo. 2s.
Rough Ways Made Smooth. Essays on Scientific Subjects. Crown 8vo. 5s. Cheap Edition, 3s. 6d.
Our Place among Infinities. Cr. 8vo. 5s.
The Expanse of Heaven. Cr. 8vo. 5s.
The Great Pyramid. Crown 8vo. 5s.
Myths and Marvels of Astronomy. Crown 8vo. 5s.
Nature Studies. By Grant Allen, A. Wilson, T. Foster, E. Clodd, and R. A. Proctor. Crown 8vo. 5s.
Leisure Readings. By E. Clodd, A. Wilson, T. Foster, A. C. Ranyard, and R. A. Proctor. Crown 8vo. 5s.

RANSOME (Cyril).—THE RISE OF CONSTITUTIONAL GOVERNMENT IN ENGLAND : being a Series of Twenty Lectures. Crown 8vo. 6s.

READER (Emily E.).—VOICES FROM FLOWER-LAND : a Birthday Book and Language of Flowers. Illustrated by ADA BROOKE. Royal 16mo. Cloth, 2s. 6d. ; vegetable vellum, 3s. 6d.

REPLY (A) TO DR. LIGHTFOOT'S ESSAYS. By the Author of 'Supernatural Religion'. 8vo. 6s.

RIBOT (Th.).—THE PSYCHOLOGY OF ATTENTION. Crown 8vo. 3s.

RICH (A.).—A DICTIONARY OF ROMAN AND GREEK ANTIQUITIES. With 2000 Woodcuts. Crown 8vo. 7s. 6d.

RICHARDSON (Dr. B. W.).—NATIONAL HEALTH. A Review of the Works of Sir Edwin Chadwick, K.C.B. Crown 4s. 6d.

RIVERS (T. and T. F.).—THE MINIATURE FRUIT GARDEN ; or, The Culture of Pyramidal and Bush Fruit Trees. With 32 Illustrations. Crown 8vo. 4s.

RIVERS (T.).—THE ROSE AMATEUR'S GUIDE. Fcp. 8vo. 4s. 6d.

ROBERTSON (A.).—THE KIDNAPPED SQUATTER, and other Australian Tales. Crown 8vo. 6s.

ROGET (John Lewis).—A HISTORY OF THE 'OLD WATER COLOUR' SOCIETY. 2 vols. Royal 8vo. 42s.

ROGET (Peter M.).—THESAURUS OF ENGLISH WORDS AND PHRASES. Crown 8vo. 10s. 6d.

ROMANES (George John, M.A., LL.D., F.R.S.).—DARWIN, AND AFTER DARWIN: an Exposition of the Darwinian Theory and a Discussion of Post-Darwinian Questions. Part I.—The Darwinian Theory. With a Portrait of Darwin and 125 Illustrations. Crown 8vo. 10s. 6d.

RONALDS (Alfred).—THE FLY-FISHER'S ETYMOLOGY. With 20 Coloured Plates. 8vo. 14s.

ROSSETTI (Maria Francesca).—A SHADOW OF DANTE: being an Essay towards studying Himself, his World, and his Pilgrimage. Cr. 8vo. 10s. 6d.

ROUND (J. H., M.A.).—GEOFFREY DE MANDEVILLE: a Study of the Anarchy. 8vo. 16s.

RUSSELL.—A LIFE OF LORD JOHN RUSSELL. By SPENCER WALPOLE. 2 vols. 8vo. 36s. Cabinet Edition, 2 vols. Crown 8vo. 12s.

SEEBOHM (Frederick).—THE OXFORD REFORMERS—JOHN COLET, ERASMUS, AND THOMAS MORE. 8vo. 14s.
——— THE ENGLISH VILLAGE COMMUNITY Examined in its Relations to the Manorial and Tribal Systems, &c. 13 Maps and Plates. 8vo. 16s.
——— THE ERA OF THE PROTESTANT REVOLUTION. With Map. Fcp. 8vo. 2s. 6d.

SEWELL (Elizabeth M.).—STORIES AND TALES. Crown 8vo. 1s. 6d. each, cloth plain; 2s. 6d. each, cloth extra, gilt edges:—

Amy Herbert.	Katharine Ashton.	Gertrude.
The Earl's Daughter.	Margaret Percival.	Ivors.
The Experience of Life.	Laneton Parsonage.	Home Life.
A Glimpse of the World.	Ursula.	After Life.
Cleve Hall.		

SHAKESPEARE.—BOWDLER'S FAMILY SHAKESPEARE. 1 vol. 8vo. With 36 Woodcuts, 14s., or in 6 vols. Fcp. 8vo. 21s.
——— OUTLINES OF THE LIFE OF SHAKESPEARE. By J. O. HALLIWELL-PHILLIPPS. With Illustrations. 2 vols. Royal 8vo. £1 1s.
——— SHAKESPEARE'S TRUE LIFE. By JAMES WALTER. With 500 Illustrations. Imp. 8vo. 21s.
——— THE SHAKESPEARE BIRTHDAY BOOK. By MARY F. DUNBAR. 32mo. 1s. 6d. cloth. With Photographs, 32mo. 5s. Drawing-Room Edition, with Photographs, Fcp. 8vo. 10s. 6d.

SHERBROOKE (Viscount).—LIFE AND LETTERS OF THE RIGHT HON. ROBERT LOWE, VISCOUNT SHERBROOKE, G.C.B. By A. PATCHETT MARTIN. With 5 Copper-plate Portraits, &c. 2 vols. 8vo.

SHIRRES (L. P.).—AN ANALYSIS OF THE IDEAS OF ECONOMICS. Crown 8vo. 6s.

SIDGWICK (Alfred).—DISTINCTION: and the Criticism of Beliefs. Cr. 8vo. 6s.

SILVER LIBRARY, The.—Crown 8vo. price 3s. 6d. each volume.

BAKER'S (Sir S. W.) Eight Years in Ceylon. With 6 Illustrations.
——— Rifle and Hound in Ceylon. With 6 Illustrations.
BARING-GOULD'S (S.) Curious Myths of the Middle Ages.
——— Origin and Development of Religious Belief. 2 vols.
BRASSEY'S (Lady) A Voyage in the 'Sunbeam'. With 66 Illustrations.

CLODD'S (E.) Story of Creation: a Plain Account of Evolution. With 77 Illustrations.
CONYBEARE (Rev. W. J.) and HOWSON'S (Very Rev. J. S.) Life and Epistles of St. Paul. 46 Illustrations.
DOUGALL'S (L.) Beggars All; a Novel.
DOYLE'S (A. Conan) Micah Clarke: a Tale of Monmouth's Rebellion.

[Continued.

SILVER LIBRARY, The.—(*Continued*).

DOYLE'S (A. Conan) The Captain of the Polestar, and other Tales.

FROUDE'S (J. A.) Short Studies on Great Subjects. 4 vols.

——— The History of England, from the Fall of Wolsey to the Defeat of the Spanish Armada. 12 vols.

——— Cæsar: a Sketch.

——— Thomas Carlyle: a History of his Life. 1795-1835. 2 vols. 1834-1881. 2 vols.

——— The Two Chiefs of Dunboy: an Irish Romance of the Last Century.

GLEIG'S (Rev. G. R.) Life of the Duke of Wellington. With Portrait.

HAGGARD'S (H. R.) She: A History of Adventure. 32 Illustrations.

——— Allan Quatermain. With 20 Illustrations.

——— Colonel Quaritch, V.C.: a Tale of Country Life.

——— Cleopatra. With 29 Full-page Illustrations.

——— Beatrice.

HARTE'S (Bret) In the Carquinez Woods, and other Stories.

HELMHOLTZ'S (Professor) Popular Lectures on Scientific Subjects. With 68 Woodcuts. 2 vols.

HOWITT'S (W.) Visits to Remarkable Places. 80 Illustrations.

JEFFERIES' (R.) The Story of My Heart. With Portrait.

——— Field and Hedgerow. Last Essays of. With Portrait.

——— Red Deer. With 17 Illust.

KNIGHT'S (E. F.) Cruise of the 'Alerte,' a Search for Treasure. With 2 Maps and 23 Illustrations.

LEES (J. A.) and CLUTTERBUCK'S (W. J.) B.C. 1887. British Columbia. 75 Illustrations.

MACAULAY'S (Lord) Essays—Lays of Ancient Rome. In 1 vol. With Portrait and Illustrations to the 'Lays'.

MACLEOD'S (H. D.) The Elements of Banking.

MARSHMAN'S (J. C.) Memoirs of Sir Henry Havelock.

MAX MÜLLER'S (F.) India, What can it teach us?

——— Introduction to the Science of Religion.

MERIVALE'S (Dean) History of the Romans under the Empire. 8 vols.

MILL'S (J. S.) Principles of Political Economy.

——— System of Logic.

MILNER'S (G.) Country Pleasures.

NEWMAN'S (Cardinal) Historical Sketches. 3 vols.

——— Fifteen Sermons Preached before the University of Oxford.

——— Apologia Pro Vita Sua.

——— Callista: a Tale of the Third Century.

——— Loss and Gain: a Tale.

——— Essays, Critical and Historical. 2 vols.

——— Sermons on Various Occasions.

——— Lectures on the Doctrine of Justification.

——— Fifteen Sermons Preached before the University of Oxford.

——— An Essay on the Development of Christian Doctrine.

——— The Arians of the Fourth Century.

——— Verses on Various Occasions.

——— Difficulties felt by Anglicans in Catholic Teaching Considered. 2 vols.

——— The Idea of a University defined and Illustrated.

——— Biblical and Ecclesiastical Miracles.

——— Discussions and Arguments on Various Subjects.

——— Grammar of Assent.

——— The Via Media of the Anglican Church. 2 vols.

——— Parochial and Plain Sermons. 8 vols.

——— Selection from 'Parochial and Plain Sermons'.

——— Discourses Addressed to Mixed Congregations.

——— Present Position of Catholics in England.

——— Sermons bearing upon Subjects of the Day.

PHILLIPPS-WOLLEY'S (C.) Snap: a Legend of the Lone Mountains. 13 Illustrations.

PROCTOR'S (R. A.) Other Worlds than Ours.

[*Continued.*

SILVER LIBRARY, The.—*(Continued.)*

PROCTOR'S (R. A.) Rough Ways made Smooth.
—— Pleasant Ways in Science.
STANLEY'S (Bishop) Familiar History of Birds. With 160 Illustrations.
STEVENSON (Robert Louis) and OSBOURNE'S (Lloyd) The Wrong Box.

WEYMAN'S (Stanley J.) The House of the Wolf: a Romance.
WOOD'S (Rev. J. G.) Petland Revisited. With 33 Illustrations.
—— 'Strange Dwellings. With 60 Illustrations.
—— Out of Doors. With 11 Illustrations.

SMITH (R. Bosworth).—CARTHAGE AND THE CARTHAGINIANS. Maps, Plans, &c. Crown 8vo. 6s.

STANLEY (E.).—A FAMILIAR HISTORY OF BIRDS. With 160 Woodcuts. Crown 8vo. 3s. 6d.

STEPHEN (Sir James).—ESSAYS IN ECCLESIASTICAL BIOGRAPHY. Crown 8vo. 7s. 6d.

STEPHENS (H. Morse).—A HISTORY OF THE FRENCH REVOLUTION. 3 vols. 8vo. Vol. I. 18s. Vol. II. 18s. [*Vol. III. in the press.*

STEVENSON (Robt. Louis).—A CHILD'S GARDEN OF VERSES. Small Fcp. 8vo. 5s.
—— A CHILD'S GARLAND OF SONGS, Gathered from 'A Child's Garden of Verses'. Set to Music by C. VILLIERS STANFORD, Mus. Doc. 4to. 2s. sewed, 3s. 6d. cloth gilt.
—— THE DYNAMITER. Fcp. 8vo. 1s. sewed, 1s. 6d. cloth.
—— STRANGE CASE OF DR. JEKYLL AND MR. HYDE. Fcp. 8vo. 1s. sewed, 1s. 6d. cloth.

STEVENSON (Robert Louis) and OSBOURNE (Lloyd).—THE WRONG BOX. Crown 8vo. 3s. 6d.

STOCK (St. George).—DEDUCTIVE LOGIC. Fcp. 8vo. 3s. 6d.

STRONG (Herbert A.), LOGEMAN (Willem S.) and WHEELER (B. I.).—INTRODUCTION TO THE STUDY OF THE HISTORY OF LANGUAGE. 8vo. 10s. 6d.

SULLY (James).—THE HUMAN MIND. 2 vols. 8vo. 21s.
—— OUTLINES OF PSYCHOLOGY. 8vo. 9s.
—— THE TEACHER'S HANDBOOK OF PSYCHOLOGY. Cr. 8vo. 5s.

SUPERNATURAL RELIGION; an Inquiry into the Reality of Divine Revelation. 3 vols. 8vo. 36s.
REPLY (A) TO DR. LIGHTFOOT'S ESSAYS. By the Author of 'Supernatural Religion'. 8vo. 7s. 6d.

SUTTNER (Bertha Von).—LAY DOWN YOUR ARMS (*Die Waffen Nieder*): The Autobiography of Martha Tilling. Translated by T. HOLMES. Crown 8vo. 7s. 6d.

SYMES (J. E.).—PRELUDE TO MODERN HISTORY: a Brief Sketch of the World's History from the Third to the Ninth Century. Cr. 8vo. 2s. 6d.

TAYLOR (Colonel Meadows).—A STUDENT'S MANUAL OF THE HISTORY OF INDIA. Crown 8vo. 7s. 6d.

THOMPSON (Annie).—A MORAL DILEMMA: a Novel. Cr. 8vo. 6s.

THOMPSON (D. Greenleaf).—THE PROBLEM OF EVIL: an Introduction to the Practical Sciences. 8vo. 10s. 6d.
—— A SYSTEM OF PSYCHOLOGY. 2 vols. 8vo. 36s.
—— THE RELIGIOUS SENTIMENTS OF THE HUMAN MIND. 8vo. 7s. 6d.
—— SOCIAL PROGRESS: an Essay. 8vo. 7s. 6d.
—— THE PHILOSOPHY OF FICTION IN LITERATURE: an Essay. Crown 8vo. 6s.

THOMSON (Most Rev. William, D.D., late Archbishop of York).—
OUTLINES OF THE NECESSARY LAWS OF THOUGHT: a Treatise on Pure and Applied Logic. Post 8vo. 6s.
THREE IN NORWAY. By Two of THEM. With a Map and 59 Illustrations. Crown 8vo. 2s. boards; 2s. 6d. cloth.

TOYNBEE (Arnold).—LECTURES ON THE INDUSTRIAL REVOLUTION OF THE 18th CENTURY IN ENGLAND. 8vo. 10s. 6d.

TREVELYAN (Sir G. O., Bart.).—THE LIFE AND LETTERS OF LORD MACAULAY.
Popular Edition. Crown 8vo. 2s. 6d. | Cabinet Edition, 2 vols. Cr. 8vo. 12s.
Student's Edition. Crown 8vo. 6s. | Library Edition, 2 vols. 8vo. 36s.
———— THE EARLY HISTORY OF CHARLES JAMES FOX. Library Edition, 8vo. 18s. Cabinet Edition, Crown 8vo. 6s.

TROLLOPE (Anthony).—THE WARDEN. Cr. 8vo. 1s. bds., 1s. 6d. cl.
———— BARCHESTER TOWERS. Crown 8vo. 1s. boards, 1s. 6d. cloth.

VERNEY (Frances Parthenope).—MEMOIRS OF THE VERNEY FAMILY DURING THE CIVIL WAR. Compiled from the Letters and Illustrated by the Portraits at Claydon House, Bucks. With 38 Portraits, Woodcuts, and Facsimile. 2 vols. Royal 8vo. 42s.

VILLE (G.).—THE PERPLEXED FARMER: How is he to meet Alien Competition? Crown 8vo. 5s.

VIRGIL.—PUBLI VERGILI MARONIS BUCOLICA, GEORGICA, ÆNEIS; the Works of VIRGIL, Latin Text, with English Commentary and Index. By B. H. KENNEDY. Crown 8vo. 10s. 6d.
———— THE ÆNEID OF VIRGIL. Translated into English Verse. By John Conington. Crown 8vo. 6s.
———— THE POEMS OF VIRGIL. Translated into English Prose. By John Conington. Crown 8vo. 6s.
———— THE ECLOGUES AND GEORGICS OF VIRGIL. Translated from the Latin by J. W. Mackail. Printed on Dutch Hand-made Paper. 16mo. 5s.
———— THE ÆNEID OF VERGIL. Books I. to VI. Translated into English Verse by JAMES RHOADES. Crown 8vo. 5s.

WAKEMAN (H. O.) and HASSALL (A.).—ESSAYS INTRODUCTORY TO THE STUDY OF ENGLISH CONSTITUTIONAL HISTORY. Edited by H. O. WAKEMAN and A. HASSALL. Crown 8vo. 6s.

WALFORD (Mrs. L. B.).—THE MISCHIEF OF MONICA. Cr. 8vo. 2s. 6d.
———— THE ONE GOOD GUEST. Crown 8vo. 6s.
———— TWELVE ENGLISH AUTHORESSES. With Portrait of HANNAH MORE. Crown 8vo. 4s. 6d.

WALKER (A. Campbell-).—THE CORRECT CARD; or, How to Play at Whist; a Whist Catechism. Fcp. 8vo. 2s. 6d.

WALPOLE (Spencer).—HISTORY OF ENGLAND FROM THE CONCLUSION OF THE GREAT WAR IN 1815 to 1858. 6 vols. Crown 8vo. 6s. each.
———— THE LAND OF HOME RULE: being an Account of the History and Institutions of the Isle of Man. Crown 8vo. 6s.

WELLINGTON.—LIFE OF THE DUKE OF WELLINGTON. By the Rev. G. R. GLEIG. Crown 8vo. 3s. 6d.

WEYMAN (Stanley J.).—THE HOUSE OF THE WOLF: a Romance. Crown 8vo. 3s. 6d.

WHATELY (Archbishop).—ELEMENTS OF LOGIC. Cr. 8vo. 4s. 6d.
———— ELEMENTS OF RHETORIC. Crown 8vo. 4s. 6d.
———— LESSONS ON REASONING. Fcp. 8vo. 1s. 6d.
———— BACON'S ESSAYS, with Annotations. 8vo. 10s. 6d.

WHISHAW (Fred. J.).—OUT OF DOORS IN TSAR LAND: a Record of the Seeings and Doings of a Wanderer in Russia. With Frontispiece and Vignette by CHARLES WHYMPER.

WILCOCKS (J. C.).—THE SEA FISHERMAN. Comprising the Chief Methods of Hook and Line Fishing in the British and other Seas, and Remarks on Nets, Boats, and Boating. Profusely Illustrated. Crown 8vo. 6s.

WILLICH (Charles M.).—POPULAR TABLES for giving Information for ascertaining the value of Lifehold, Leasehold, and Church Property, the Public Funds, &c. Edited by H. BENCE JONES. Crown 8vo. 10s. 6d.

WITT (Prof.)—Works by. Translated by Frances Younghusband.
———— THE TROJAN WAR. Crown 8vo. 2s.
———— MYTHS OF HELLAS; or, Greek Tales. Crown 8vo. 3s. 6d.
———— THE WANDERINGS OF ULYSSES. Crown 8vo. 3s. 6d.
———— THE RETREAT OF THE TEN THOUSAND; being the Story of Xenophon's 'Anabasis'. With Illustrations. Crown 8vo. 3s. 6d.

WOLFF (Henry W.).—RAMBLES IN THE BLACK FOREST. Crown 8vo. 7s. 6d.
———— THE WATERING PLACES OF THE VOSGES. With Map. Crown 8vo. 4s. 6d.
———— THE COUNTRY OF THE VOSGES. With a Map. 8vo. 12s.
———— PEOPLE'S BANKS: a Record of Social and Economic Success. 8vo. 7s. 6d.

WOOD (Rev. J. G.).—HOMES WITHOUT HANDS; a Description of the Habitations of Animals. With 140 Illustrations. 8vo. 7s. net.
———— INSECTS AT HOME; a Popular Account of British Insects, their Structure, Habits, and Transformations. With 700 Illustrations. 8vo. 7s. net.
———— INSECTS ABROAD; a Popular Account of Foreign Insects, their Structure, Habits, and Transformations. With 600 Illustrations. 8vo. 7s. net.
———— BIBLE ANIMALS; a Description of every Living Creature mentioned in the Scriptures. With 112 Illustrations. 8vo. 7s. net.
———— STRANGE DWELLINGS; abridged from 'Homes without Hands'. With 60 Illustrations. Crown 8vo. 3s. 6d.
———— OUT OF DOORS; a Selection of Original Articles on Practical Natural History. With 11 Illustrations. Crown 8vo. 3s. 6d.
———— PETLAND REVISITED. With 33 Illustrations. Crown 8vo. 3s. 6d.

WORDSWORTH (Bishop Charles).—ANNALS OF MY LIFE. First Series, 1806-1846. 8vo. 15s. Second Series, 1847-1856. 8vo.

WYLIE (J. H.).—HISTORY OF ENGLAND UNDER HENRY THE FOURTH. Crown 8vo. Vol. I. 10s. 6d.; Vol. II.

ZELLER (Dr. E.).—HISTORY OF ECLECTICISM IN GREEK PHILOSOPHY. Translated by Sarah F. Alleyne. Crown 8vo. 10s. 6d.
———— THE STOICS, EPICUREANS, AND SCEPTICS. Translated by the Rev. O. J. Reichel. Crown 8vo. 15s.
———— SOCRATES AND THE SOCRATIC SCHOOLS. Translated by the Rev. O. J. Reichel. Crown 8vo. 10s. 6d.
———— PLATO AND THE OLDER ACADEMY. Translated by Sarah F. Alleyne and Alfred Goodwin. Crown 8vo. 18s.
———— THE PRE-SOCRATIC SCHOOLS. Translated by Sarah F. Alleyne. 2 vols. Crown 8vo. 30s.
———— OUTLINES OF THE HISTORY OF GREEK PHILOSOPHY. Translated by Sarah F. Alleyne and Evelyn Abbott. Crown 8vo. 10s. 6d.

www.ingramcontent.com/pod-product-compliance
Lightning Source LLC
Chambersburg PA
CBHW020319240426

43673CB00039B/863